"*Redeeming Mathematics* is a valuable addition the growing literature on the relationship between mathematics and Christian belief. Poythress's treatment of three distinct dimensions of mathematics—as transcendent abstract truths, as part of the physical world, and as comprehensible to human beings—is a unique and helpful addition to the conversation on this relationship. The book is accessible to nonspecialists, but even those who are well-versed in these matters will find much to interest and challenge them."

James Bradley, Professor Emeritus of Mathematics, Calvin College; author, *Mathematics Through the Eyes of Faith*; Editor, *Journal of the Association of Christians in the Mathematical Sciences*

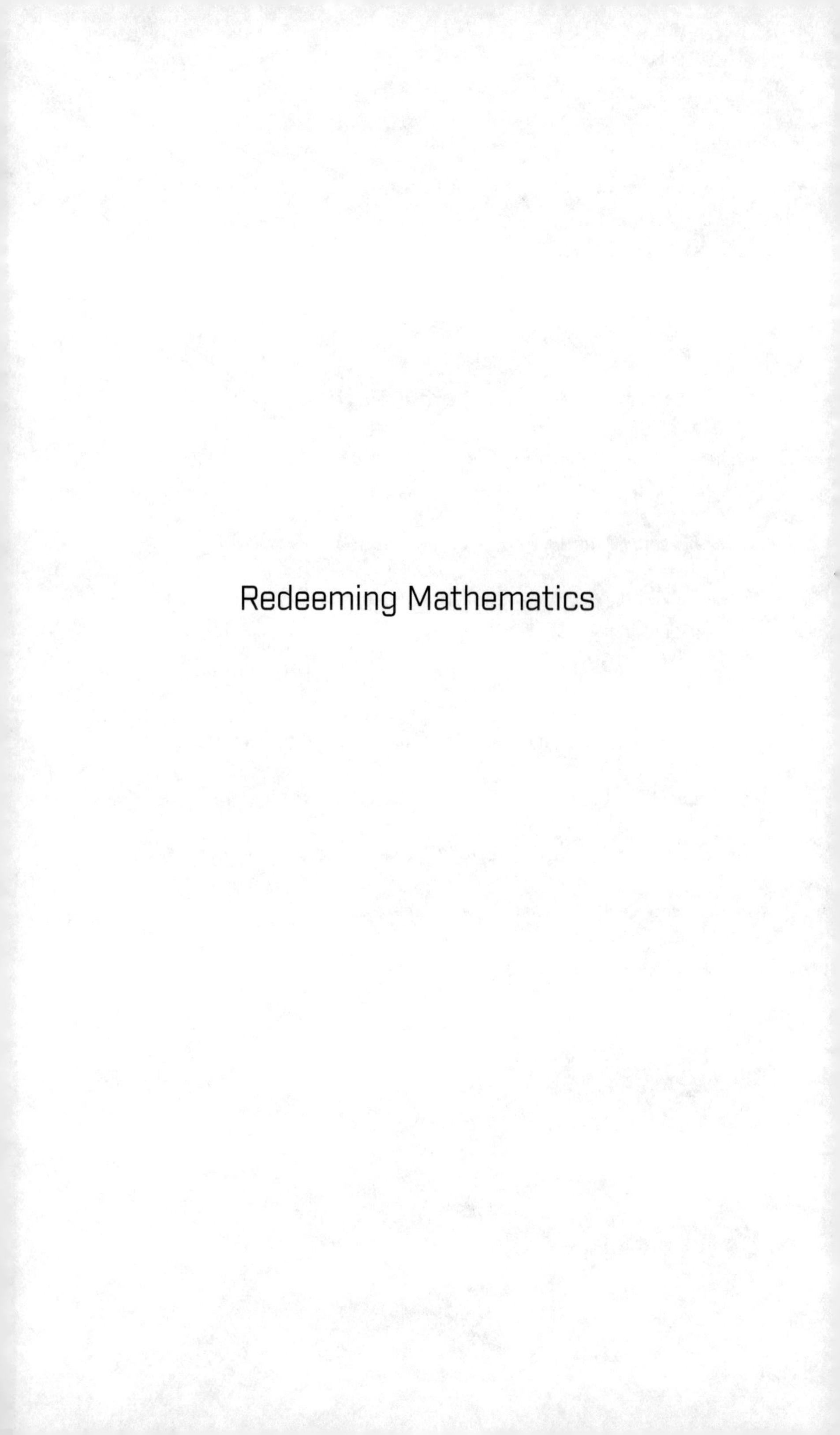

Redeeming Mathematics

Other Crossway Books by Vern S. Poythress

Chance and the Sovereignty of God: A God-Centered Approach to Probability and Random Events

In the Beginning Was the Word: Language—A God-Centered Approach

Inerrancy and the Gospels: A God-Centered Approach to the Challenges of Harmonization

Inerrancy and Worldview: Answering Modern Challenges to the Bible

Logic: A God-Centered Approach to the Foundation of Western Thought

Redeeming Philosophy: A God-Centered Approach to the Big Questions

Redeeming Science: A God-Centered Approach

Redeeming Sociology: A God-Centered Approach

Redeeming Mathematics

A God-Centered Approach

Vern S. Poythress

WHEATON, ILLINOIS

Redeeming Mathematics: A God-Centered Approach

Published by Crossway
1300 Crescent Street
Wheaton, Illinois 60187

Cover design: Matt Naylor

First printing 2015

Printed in the United States of America

All emphases in Scripture quotations have been added by the author.

Trade paperback ISBN: 978-1-4335-4110-0
ePub ISBN: 978-1-4335-4113-1
PDF ISBN: 978-1-4335-4111-7
Mobipocket ISBN: 978-1-4335-4112-4

Library of Congress Cataloging-in-Publication Data

Poythress, Vern S.
Redeeming mathematics : a God-centered approach / Vern S. Poythress.
pages cm.
Includes bibliographical references and index.
ISBN 978-1-4335-4110-0 (trade paperback)
1. Bible—Evidences, authority, etc. 2. Mathematics—Religious aspects—Christianity. 3. Mathematics in the Bible. I. Title.
BS540.P69 2015
230—dc23 2014015579

Crossway is a publishing ministry of Good News Publishers.

5L 28 27 26 25 24 23 22 21 20
14 13 12 11 10 9 8 7 6 5 4 3 2

Contents

Part IV

Other Kinds of Numbers

Part V

Geometry and Higher Mathematics

Supplements

Diagrams and Illustrations

Diagrams

Illustrations

Introduction

Why God?

Does God have anything to do with mathematics? Many people have never considered the question. It seems to them that the truths of mathematics are just "out there."[1] In their view, mathematics presents us with a world remote from religious questions. Some people think that God exists; others are convinced that he does not; still others would say that they do not know. But all of them might say, "It does not matter when we look at mathematics."

I think it *does* matter. In this book I intend to show why. I am working from the conviction that we should honor and glorify God in all of life: "So, whether you eat or drink, or whatever you do, do all to the glory of God" (1 Cor. 10:31). The expression "whatever you do" includes our *thinking*, and our thinking includes our thinking about mathematics. In addition, I am a follower of Christ, and I acknowledge that Christ is Lord of all.[2] If he is Lord of all, he is also Lord of mathematics. But what does that mean? We will try to work out the implications.

I am writing primarily to people who follow Christ, who have come to know him as the living Savior and who have put their faith in him. They find out from the Bible that Christ himself teaches that the Old Testament

---

[1] Other people think that arithmetic truths are "in here," that is, that they are items of mental furniture. We certainly do have mental concepts concerning mathematics. But, as we shall see later, mathematics ought not to be reduced to this pole of subjective experience.

[2] I have been encouraged here by Abraham Kuyper, who challenged people to think about the universal lordship of Christ in *Lectures on Calvinism: Six Lectures Delivered at Princeton University under Auspices of the L. P. Stone Foundation* (Grand Rapids, MI: Eerdmans, 1931). See Vern S. Poythress, *Redeeming Philosophy: A God-Centered Approach to the Big Questions* (Wheaton, IL: Crossway, 2014), appendix A.

is the word of God, God's own speech to us in written form (see especially Matt. 5:17–18; 19:4–5; John 10:35). The Old Testament predicts the coming of Christ (see, for example, Isa. 9:6–7; 11:1–5; 53:1–12; Mic. 5:2). It also makes provision for later prophets (Deut. 18:15–22). After Christ completed his work on earth, the New Testament was written with the same authority as the Old Testament. So I am going to draw on the Bible for understanding who God is, and in addition for understanding what mathematics is.[3]

If you are not yet a follower of Christ, you are still welcome to read. I hope it will be informative for you to learn what are the implications of the Bible for mathematics. But if you are going to appropriate the truth for yourself, you will first of all have to come to terms with Christ. You should ask who he is and what he has to say about you and the way you live your life. I would recommend that you start by reading the part of the Bible consisting in the Gospels (Matthew, Mark, Luke, and John).

[3] For extended discussion of the nature of the Bible, many books are available. See especially John M. Frame, *The Doctrine of the Word of God* (Phillipsburg, NJ: Presbyterian & Reformed, 2010). For a discussion of the broader set of commitments with which to study the Bible, see Poythress, *Redeeming Philosophy*; and Vern S. Poythress, *Inerrancy and Worldview: Answering Modern Challenges to the Bible* (Wheaton, IL: Crossway, 2012).

Part I

Basic Questions

1

God and Mathematics

Let us begin with numbers. We can consider a particular case: 2 + 2 = 4. That is true. It was true yesterday. And it always will be true. It is true everywhere in the universe. We do not have to travel out to distant galaxies to check it. Why not? We just know. Why do we have this conviction? Is it not strange? What is it about 2 + 2 = 4 that results in this conviction about its universal truth?[1]

All Times and All Places

2 + 2 = 4 is true at all times and at all places.[2] We have classic terms to describe this situation: the truth is omnipresent (present at all places) and eternal (there at all times). The truth 2 + 2 = 4 has these two characteristics or *attributes* that are classically attributed to God. So is God in our picture, already at this point? We will see.

Technically, God's eternity is usually conceived of as being "above" or "beyond" time. But words like "above" and "beyond" are metaphorical and point to mysteries. There is, in fact, an analogous mystery with respect to 2 + 2 = 4. If 2 + 2 = 4 is universally true, is it not in some sense "beyond" the particularities of any one place or time?

Moreover, the Bible indicates that God is not only "above" time in the

[1] Some relativists and multiculturalists might claim that even the truth of 2 + 2 = 4 is "relative" to culture. But in their practical living they show that they are confident about such truths.
[2] The subsequent analysis of the truth borrows ideas and wording from Vern S. Poythress, *Redeeming Science: A God-Centered Approach* (Wheaton, IL: Crossway, 2006), chapters 1 and 14.

sense of not being subject to the limitations of finite creaturely experience of time, but is "in" time in the sense of acting in time and interacting with his creatures.[3] Similarly, 2 + 2 = 4 is "above" time in its universality, but "in" time through its applicability to each particular situation. Two apples plus two more apples is four apples.

Divine Attributes of Arithmetical Truth

The attributes of omnipresence and eternity are only the beginning. On close examination, other divine attributes seem to belong to arithmetical truths.

Consider. If 2 + 2 = 4 holds for all times, we are presupposing that it is the *same* truth through all times. The truth does not change with time. It is immutable.

Next, 2 + 2 = 4 is at bottom ideational in character. We do not literally see the truth 2 + 2 = 4, but only particular instances to which it applies: two apples plus two apples. The truth that 2 + 2 = 4 is essentially immaterial and invisible, but is known through manifestations. Likewise, God is essentially immaterial and invisible, but is known through his acts in the world.

Next, we have already observed that 2 + 2 = 4 is true. Truthfulness is also an attribute of God.

The Power of Arithmetical Truth

Next, consider the attribute of power. Mathematicians make their formulations to describe properties of numbers. The properties are there before the mathematicians make their formulations. The human mathematical formulation follows the facts and is dependent on them. An arithmetical truth or regularity must hold for a whole series of cases. The mathematician cannot force the issue by inventing a new property, say that 2 + 2 = 5, and then forcing the universe to conform to his formulation. (Of course, the written symbols such as *4* and *5* that denote the numbers could have been chosen differently. And a mathematician can define a

[3] John M. Frame, *The Doctrine of God* (Phillipsburg, NJ: Presbyterian & Reformed, 2002), 543–575.

new abstract object to have properties that he chooses. But we do not "choose" the properties of natural numbers.) Natural numbers conform to arithmetical properties and laws that are already there, laws that are discovered rather than invented. The laws must already be there. 2 + 2 = 4 must actually hold. It must "have teeth." If it is truly universal, it is not violated. Two apples and two apples always make four apples. No event escapes the "hold" or dominion of arithmetical laws. The power of these laws is absolute, in fact, infinite. In classical language, the law is omnipotent ("all powerful").

2 + 2 = 4 is both transcendent and immanent. It transcends the creatures of the world by exercising power over them, conforming them to its dictates. It is immanent in that it touches and holds in its dominion even the smallest bits of this world.[4] 2 + 2 = 4 transcends the galactic clusters and is immanently present in the behavior of the electrons surrounding a beryllium nucleus. Transcendence and immanence are characteristics of God.

The Personal Character of Law

Many agnostics and atheists by this time will be looking for a way of escape. It seems that the key concept of arithmetical truth is beginning to look suspiciously like the biblical idea of God. The most obvious escape, and the one that has rescued many from spiritual discomfort, is to deny that arithmetical truth is personal. It is just there as an impersonal something.

Throughout the ages people have tried such routes. They have constructed idols, substitutes for God. In ancient times, the idols often had the form of statues representing a god—Poseidon, the god of the sea, or Mars, the god of war. Nowadays in the Western world we are more sophisticated. Idols now take the form of mental constructions of a god or a God-substitute. Money and pleasure can become idols. So can "humanity" or "nature" when it receives a person's ultimate allegiance. "Scientific

[4] On the biblical view of transcendence and immanence, see John M. Frame, *The Doctrine of the Knowledge of God* (Phillipsburg, NJ: Presbyterian & Reformed, 1987), especially 13–15; and Frame, *Doctrine of God*, especially 107–115. On the relationship to cosmonomic philosophy, see Poythress, *Redeeming Philosophy*, appendix A.

law," when it is viewed as impersonal, becomes another God-substitute. Arithmetical truth, as a particular kind of scientific law, is also viewed as impersonal. In both ancient times and today, idols conform to the imagination of the one who makes them. Idols have enough similarities to the true God to be plausible, but differ so as to allow us comfort and the satisfaction of manipulating the substitutes that we construct.

In fact, however, a close look at 2 + 2 = 4 shows that this escape route is not really plausible. Law implies a law-giver. Someone must think the law and enforce it, if it is to be effective. But if some people resist this direct move to personality, we may move more indirectly.

Scientists and mathematicians in practice believe passionately in the rationality of scientific laws and arithmetical laws. We are not dealing with something totally irrational, unaccountable, and unanalyzable, but with lawfulness that in some sense is accessible to human understanding. Rationality is a sine qua non for scientific law. But, as we know, rationality belongs to persons, not to rocks, trees, and subpersonal creatures. If the law is rational, as mathematicians assume it is, then it is also personal.

Scientists and mathematicians also assume that laws can be articulated, expressed, communicated, and understood through human language. Mathematical work includes not only rational thought but symbolic communication. Now, the original law, the law 2 + 2 = 4 that is "out there," is not known to be written or uttered in a human language. But it must be expressible in language in our secondary description. It must be translatable into not only one but many human languages. We may explain the meaning of the symbols and the significance and application of 2 + 2 = 4 through clauses, phrases, explanatory paragraphs, and contextual explanations in human language.

Arithmetical laws are clearly like human utterances in their ability to be grammatically articulated, paraphrased, translated, and illustrated.[5] Law is utterance-like, language-like. And the complexity of utterances that we find among mathematicians, as well as among human beings in general, is not duplicated in the animal world.[6] Language is one of the

[5] Vern S. Poythress, "Tagmemic Analysis of Elementary Algebra," *Semiotica* 17/2 (1976): 131–151.

[6] Animal calls and signals do mimic certain limited aspects of human language. And chimpanzees can be taught to respond to symbols with meaning. But this is still a long way from the complex grammar and mean-

defining characteristics that separates man from animals. Language, like rationality, belongs to persons. It follows that arithmetical laws are in essence personal.[7]

The Incomprehensibility of Law

In addition, law is both knowable and incomprehensible in the theological sense. That is, we know arithmetical truths, but in the midst of this knowledge there remain unfathomed depths and unanswered questions about the very areas where we know the most. Why does $2 + 2 = 4$ hold everywhere?

The knowability of laws is closely related to their rationality and their immanence, displayed in the accessibility of effects. We experience incomprehensibility in the fact that the increase of mathematical understanding only leads to ever deeper questions: "How can this be?" and "Why this law rather than many other ways that the human mind can imagine?" The profundity and mystery in mathematical discoveries can only produce awe—yes, worship—if we have not blunted our perception with hubris (Isa. 6:9–10).

Are We Divinizing Nature?

But now we must consider an objection. By claiming that arithmetical laws have divine attributes, are we divinizing nature? That is, are we taking something out of the created world and falsely claiming that it is divine? Are not arithmetical laws a part of the created world? Should we not classify them as creature rather than Creator?[8]

I suspect that the specificity of arithmetical laws, their obvious

---

ing of human language. See, e.g., Stephen R. Anderson, *Doctor Dolittle's Delusion: Animals and the Uniqueness of Human Language* (New Haven, CT: Yale University Press, 2004).

[7] In their ability to undergo transformation and reformulation, scientific laws also show an analogy with the ability of human language to represent multiple perspectives. For more on the language-like character of scientific law and mathematics, see Vern S. Poythress, "Science as Allegory," *Journal of the American Scientific Affiliation* 35/2 (1983): 65–71, http://www.frame-poythress.org/science-as-allegory/, accessed June 18, 2014; Vern S. Poythress, "Newton's Laws as Allegory," *Journal of the American Scientific Affiliation* 35/3 (1983): 156–161, http://www.frame-poythress.org/newtons-laws-as-allegory/, accessed June 18, 2014; Vern S. Poythress, "Mathematics as Rhyme," *Journal of the American Scientific Affiliation* 35/4 (1983): 196–203, http://www.frame-poythress.org/mathematics-as-rhyme/, accessed June 18, 2014.

[8] In conformity with the Bible (especially Genesis 1), we maintain that God and the created world are distinct. God is not to be identified with the creation or any part of it, nor is the creation a "part" of God. The Bible repudiates all forms of pantheism and panentheism.

reference to the created world, has become the occasion for many of us to infer that these laws are a part of the created world. But such an inference is clearly invalid. The speech describing a butterfly is not itself a butterfly or a part of a butterfly. Speech referring to the created world is not necessarily an ontological part of the world to which it refers.

The Bible indicates that God rules the world through his speech.[9] He speaks, and it is done:

> By the *word* of the Lord the heavens were made,
> and by the *breath of his mouth* all their host. (Ps. 33:6)

> For he *spoke*, and it came to be;
> he *commanded*, and it stood firm. (Ps. 33:9)

> And God *said*, "Let there be light," and there was light. (Gen. 1:3)

God also continually sustains the world by his word: "he upholds the universe by the *word* of his power" (Heb. 1:3). God's word has divine wisdom, power, truth, and holiness. It has *divine* attributes, because it expresses God's own character. God expresses rather than undermines his own deity when he speaks words that address the created world.

We may then conclude that the same principle applies in particular to numerical truths about the world. God governs *everything*, including numerical truth. His word specifies what is true. The apples in a group of four apples are created things. What God says about them is divine. In other words, his word specifies that $2 + 2 = 4$.

The key idea that the law for the world is divine is even older than the rise of Christianity. Even before the coming of Christ people noticed profound regularity in the government of the world and wrestled with the meaning of this regularity. Both the Greeks (especially the Stoics) and the Jews (especially Philo) developed speculations about the logos, the divine "word" or "reason" behind what is observed.[10] In addition the Jews had the Old Testament, which reveals the role of the word of God in creation and providence. Against this background John 1:1 proclaims, "In

[9] See the discussion in Poythress, *Redeeming Science*, chapter 1.

[10] See R. B. Edwards, "Word," in Geoffrey W. Bromiley et al., eds., *The International Standard Bible Encyclopedia*, 4 vols. (Grand Rapids, MI: Eerdmans, 1988), 4:1103–1107, and the associated literature.

the beginning was the Word, and the Word was with God, and the Word was God." John responds to the speculations of his time with a striking revelation: that the Word (logos) that created and sustains the universe is not only a divine person "with God," but the very One who became incarnate: "the Word became flesh" (1:14).

God said, "Let there be light" (Gen. 1:3). He referred to light as a part of the created world. But precisely in this reference, his word has divine power to bring creation into being. The effect in creation took place at a particular time. But the plan for creation, as exhibited in God's word, is eternal. Likewise, God's speech to us in the Bible refers to various parts of the created world, but the speech (in distinction to the things to which it refers) is divine in power, authority, majesty, righteousness, eternity, and truth.[11]

The analogy with the incarnation should give us our clue. The second person of the Trinity, the eternal Word of God, became man in the incarnation but did not therefore cease to be God. Likewise, when God speaks and says what is to be the case in this world, his words do not cease to have the divine power and unchangeability that belong to him. Rather, they remain divine and in addition have the power to specify the situation with respect to creaturely affairs. God's word remains divine when it becomes law, a specific directive with respect to this created world. In particular, 2 + 2 = 4 remains a divinely ordained truth when it becomes a specific directive with respect to four apples on the kitchen table.

The Goodness of Law

Is 2 + 2 = 4 morally good? An arithmetical truth is not directly a moral precept. But indirectly it requires us to conform to it. We have an ethical constraint to believe the truth, once we have become convinced of it. We can also say that in a wider sense it is "good" for the universe and for us that 2 + 2 = 4. It never lies. We would not be able to live, nor would the universe hold together, without the consistency of arithmetical truths.

[11] On the divine character of God's word, see Vern S. Poythress, *God-Centered Biblical Interpretation* (Phillipsburg, NJ: Presbyterian & Reformed, 1999), 32–36.

The Beauty of Law

Is 2 + 2 = 4 beautiful? I think so. But not everyone is good at seeing the beauty in mathematics. I think there is beauty in the simplicity of 2 + 2 = 4. It is in harmony with the world. It is beautiful that its truth is displayed repeatedly, in four apples, four pencils, and four chairs. It is beautiful in its harmony with other arithmetical truths, with which it can be combined.

The beauty in arithmetic shows the beauty of God himself. Though beauty has not been a favorite topic in classical expositions of the doctrine of God, the Bible shows us a God who is profoundly beautiful. He manifests himself in beauty in the design of the tabernacle, the poetry of the Psalms, and the elegance of Christ's parables, as well as the moral beauty of the life of Christ.

The beauty of God himself is reflected in what he has made. We are accustomed to seeing beauty in particular objects within creation, such as a butterfly or a lofty mountain or a flower-covered meadow. But beauty is also displayed in the simple, elegant form of some of the most basic physical laws, like Newton's law for force, F = ma, or Einstein's formula relating mass and energy, $E = mc^2$. The same goes for the simple beauties in arithmetic and the more profound beauties that mathematicians discover in advanced mathematics.

The Rectitude of 2 + 2 = 4

Another attribute of God is righteousness. God's righteousness is displayed preeminently in the moral law and in the moral rectitude of his judgments, that is, his rewards and punishments based on moral law. Does God's rectitude appear in mathematics? The traces are somewhat less obvious, but still present. People could try to disobey arithmetical laws, for example, when they are trying to balance their checkbook. If they do, they may suffer for it. There is a kind of built-in righteousness in the way in which arithmetical laws lead to consequences.

In addition, the rectitude of God is closely related to the fitness of his acts. It fits the character of who God is that we should worship him alone (Ex. 20:3). It fits the character of human beings made in the image of God that they should imitate God by keeping the Sabbath (vv. 8–11). Human actions fitly correspond to the actions of God.

In addition, punishments must be fitting. Death is the fitting or matching penalty for murder (Gen. 9:6). "As you have done, it shall be done to you; your deeds shall return on your own head" (Obadiah 15). The punishment fits the crime. There is a symmetrical match between the nature of the crime and the punishment that fits it.[12] In the arena of arithmetical law we do not deal with crimes and punishments. But rectitude expresses itself in symmetries, in orderliness, in a "fittingness" to the character of arithmetic. This "fitness" is perhaps closely related to beauty. God's attributes are involved in one another and imply one another, so beauty and righteousness are closely related. It is the same with the area of arithmetical law. Arithmetical laws are both beautiful and "fitting," demonstrating rectitude.

Law as Trinitarian

Does 2 + 2 = 4 specifically reflect the Trinitarian character of God? Philosophers have sometimes maintained that one can infer the existence of God, but not the Trinitarian character of God, on the basis of the world around us. Romans 1:18–21 indicates that unbelievers know God, but how much do they know? I am not addressing this difficult question, but rather reflecting on what we can discern about the world once we have absorbed biblical teaching about God.

God has specified by his word that 2 + 2 = 4. Thus, in its origin the truth that 2 + 2 = 4 is a form of the word of God. Hence, it reflects the Trinitarian statement in John 1:1, which identifies the second person of the Trinity as the eternal Word. In John, God the Father is the speaker of the Word, and God the Son is the Word who is spoken. John 1 does not explicitly mention the Holy Spirit. But earlier Scriptures associate the Spirit with the "breath" of God that carries the word out.

"By the word of the Lord the heavens were made, and by the *breath of his mouth* all their host" (Ps. 33:6). The Hebrew word here for breath is

[12] See the extended discussion of just punishment in Vern S. Poythress, *The Shadow of Christ in the Law of Moses* (Phillipsburg, NJ: Presbyterian & Reformed, 1995), 119–249.

ruach, the same word that is regularly used for the Holy Spirit. Indeed, the designation of the third person of the Trinity as "Spirit" (Hebrew *ruach*) already suggests the association that becomes more explicit in Psalm 33:6. Similarly, Ezekiel 37 evokes three different meanings of the Hebrew word *ruach*, namely "breath" (vv. 5, 10), "winds" (v. 9), and "Spirit" (v. 14). The vision in Ezekiel 37 clearly represents the Holy Spirit as the breath of God coming into human beings to give them life. Thus all three persons of the Trinity are present in distinct ways when God speaks his Word. The three persons are therefore all present in God's speech specifying that 2 + 2 = 4.

We can come at the issue another way. Dorothy Sayers acutely observes that the experience of a human author writing a book contains profound analogies to the Trinitarian character of God.[13] An author's act of creation in writing imitates the action of God in creating the world. God creates according to his Trinitarian nature. A human author creates with an Idea, Energy, and Power, corresponding mysteriously to the involvement of the three persons in creation. Without tracing Sayers's reflections in detail, we may observe that the act of God in creation does involve all three persons. God the Father is the originator. God the Son, as the eternal Word (John 1:1–3), is involved in the words of command that issue from God ("Let there be light"; Gen. 1:3). God the Spirit hovers over the waters (v. 2). Psalm 104:30 says that "when you send forth your Spirit, they [animals] are created." Moreover, the creation of Adam involves an inbreathing by God that alludes to the presence of the Spirit (Gen. 2:7). Though the relation among the persons of the Trinity is deeply mysterious, and though all persons are involved in all the actions of God toward the world, one can distinguish different aspects of action belonging preeminently to the different persons.

2 + 2 = 4 stems from the activity of God, the "Author" of creation. The activity of all three persons is therefore implicit in the very nature of the truth 2 + 2 = 4. First, 2 + 2 = 4 involves a rationality that implies the coherence of thought. This corresponds to Sayers's term "Idea," representing the plan of the Father. Second, in its application to the world, 2 + 2 = 4 involves an articulation, a specification, an expression of the plan, with

[13] Dorothy Sayers, *The Mind of the Maker* (New York: Harcourt, Brace, 1941), especially 33–46.

respect to all the particulars of a world. This specification corresponds to Sayers's term "Energy" or "Activity," representing the Word, who is the expression of the Father. Third, the expression of the truth that 2 + 2 = 4 involves holding created things responsible to its truth: it involves a concrete application to creatures, bringing them to respond to the law as willed by the Father. This corresponds to Sayers's term "Power," representing the Spirit.[14]

God Showing Himself

These relations are suggestive, but we need not develop the thinking further at this point. It suffices to observe that, in reality, the word specifying that 2 + 2 = 4 is divine. We are speaking of God himself and his revelation of himself through his governance of the world. People working with mathematics rely on God's word in order to carry out their work. When we analyze what 2 + 2 = 4 really is, we find that arithmetic constantly confronts us with God himself, the Trinitarian God; we are constantly depending on who he is and what he does in conformity with his divine nature. In thinking about arithmetic, we are thinking God's thoughts after him.[15]

But Do People Who Calculate Believe?

But do people who work with numbers really believe all this? They do and they do not. The situation has already been described in the Bible:

> For what can be known about God is plain to them, because God has shown it to them. For his invisible attributes, namely, his eternal power and divine nature, have been clearly perceived, ever since the creation of the world, in the things that have been made. So they are without excuse. (Rom. 1:19–20)

> The heavens declare the glory of God,
>   and the sky above proclaims his handiwork.

[14] See also John Milbank, *The Word Made Strange: Theology, Language, Culture* (Oxford: Blackwell, 1997), on the Trinitarian roots of communication.
[15] See Poythress, *God-Centered Biblical Interpretation*, 31–50.

> Day to day pours out speech,
> and night to night reveals knowledge. (Ps. 19:1–2)

They know God. They rely on him. But because this knowledge is morally and spiritually painful, they also suppress and distort it:

> For although they knew God, they did not honor him as God or give thanks to him, but they became futile in their thinking, and their foolish hearts were darkened. Claiming to be wise, they became fools, and exchanged the glory of the immortal God for images resembling mortal man and birds and animals and creeping things. (Rom. 1:21–23)

Modern people may no longer make idols in the form of physical images, but their very idea of arithmetical laws is an idolatrous twisting of their knowledge of God. They conceal from themselves the fact that the "law" is personal and that they are responsible to the Person.

Even in their rebellion, people continue to depend on God being there. They show in action that they continue to believe in God. Cornelius Van Til compares it to an incident he saw on a train, where a small girl sitting on her father's lap slapped him in the face.[16] The rebel must depend on God, and must be "sitting on his lap," even to be able to engage in rebellion.

Do We Christians Believe?

The fault, I suspect, is not entirely on the side of unbelievers. The fault also occurs among Christians. Christians have sometimes adopted an unbiblical concept of God that moves him one step out of the way of our ordinary affairs. We ourselves may think of "scientific law" or "natural law" or mathematics as a kind of cosmic mechanism or impersonal clockwork that runs the world most of the time, while God is on vacation. God comes and acts only rarely through miracle. But this is not biblical. "You

[16] Cornelius Van Til, "Transcendent Critique of Theoretical Thought" (Response by C. Van Til), in *Jerusalem and Athens: Critical Discussions on the Theology and Apologetics of Cornelius Van Til*, ed. E. R. Geehan (n.l.: Presbyterian & Reformed, 1971), 98. For rebels' dependence on God, see Cornelius Van Til, *The Defense of the Faith*, 2nd ed. (Philadelphia: Presbyterian & Reformed, 1963); and the exposition by John M. Frame, *Apologetics to the Glory of God: An Introduction* (Phillipsburg, NJ: Presbyterian & Reformed, 1994).

cause the grass to grow for the livestock" (Ps. 104:14). "He gives snow like wool" (Ps. 147:16). Let us not forget it. If we ourselves recovered a robust doctrine of God's involvement in daily caring for his world in detail, we would find ourselves in a much better position to dialogue with atheists who rely on that same care.

Principles for Witness

In order to use this situation as a starting point for witness, we need to bear in mind several principles.

First, the observation that God underlies the truth 2 + 2 = 4 does not have the same shape as the traditional theistic proofs—at least as they are often understood. We are not trying to lead people to come to know a God who is completely new to them. Rather, we show that they already know God as an aspect of their human experience. This places the focus not on intellectual debate but on being a full human being.[17]

Second, people deny God within the very same context in which they depend on him. The denial of God springs ultimately not from intellectual flaws or from failure to see all the way to the conclusion of a chain of syllogistic reasoning, but from spiritual failure. We are rebels against God, and we will not serve him. Consequently, we suffer under his wrath (Rom. 1:18), which has intellectual as well as spiritual and moral effects. Those who rebel against God are "fools," according to Romans 1:22.

Third, it is humiliating to intellectuals to be exposed as fools, and it is further humiliating, even psychologically unbearable, to be exposed as guilty of rebellion against the goodness of God. We can expect our hearers to fight with a tremendous outpouring of intellectual and spiritual energy against so unbearable an outcome.

Fourth, the gospel itself, with its message of forgiveness and reconciliation through Christ, offers the only remedy that can truly end this fight against God. But it brings with it the ultimate humiliation: that my restoration comes entirely from God, from outside me—in spite of, rather

[17] Much valuable insight into the foundations of apologetics is to be found in the tradition of transcendental apologetics founded by Cornelius Van Til. See Van Til, *Defense of the Faith*; and Frame, *Apologetics to the Glory of God*.

than because of, my vaunted abilities. To climax it all, so wicked was I that it took the price of the death of the Son of God to accomplish my rescue.

Fifth, approaching people in this way constitutes spiritual warfare. Unbelievers and idolaters are captives to Satanic deceit (1 Cor. 10:20; Eph. 4:17–24; 2 Thess. 2:9–12; 2 Tim. 2:25–26; Rev. 12:9). They do not get free from Satan's captivity unless God gives them release (2 Tim. 2:25–26). We must pray to God and rely on God's power rather than the ingenuity of human argument and eloquence of persuasion (1 Cor. 2:1–5; 2 Cor. 10:3–5).

Sixth, we come into this encounter as fellow sinners. Christians too have become massively guilty by being captive to the idolatry in which scientific and arithmetical law is regarded as impersonal. Within this captivity we take for granted the benefits and beauties of science and mathematics for which we should be filled with gratitude and praise to God.

Does an approach to witnessing based on these principles work itself out differently from many of the approaches that attempt to address intellectuals? To me it appears so.

2

The One and the Many

Numbers are related to an old philosophical problem, called the problem of the one and the many. We can also describe it as the problem of unity and diversity. How do unity and diversity fit together? It is worthwhile understanding a little about the problem.[1]

The Philosophical Problem of One and Many

Philosophers in ancient Greece already confronted the problem. How does the multiplicity of things that we observe relate to the unity of *one* world and the unity belonging to every member of a particular class? How does the unity of the class of cats relate to the particularity of Felix the cat and each other cat? Parmenides and later Plotinus said that the one was prior to the many.[2] But if we start with one thing, and it has no differentiation, how can it differentiate later or lead to the observed differences among things in the world? Heraclitus and the atomists said that the many were prior to the one. But if we start with many things, how can they then be related to one another, and why do they exhibit the common characteristics of belonging to one class (like the class of cats)?

[1] See Cornelius Van Til, *The Defense of the Faith*, 2nd ed., rev. and abridged (Philadelphia: Presbyterian & Reformed, 1963), 25–26; Vern S. Poythress, *Logic: A God-Centered Approach to the Foundation of Western Thought* (Wheaton, IL: Crossway, 2013), chapter 18.

[2] I simplify. Parmenides and Heraclitus are widely known for their contrasting positions on the nature of change. Parmenides said that what was real never changed, and thus change was an illusion. Heraclitus said that everything changed. These two positions exhibit the problem of the one and the many within the framework of time. Later philosophers focused on the problem of the one and the many as exhibited in classes of things. What is common to everything in a class is the one; the many members of the class are the many. Which is logically prior, catness (the one) or a plurality of cats (the many)?

Medieval philosophy continued to consider the question. On one side of the dispute were philosophical *realists*. These people said that universal categories like the category *cat* or *horse* were *real*. (This kind of *realism* should not be confused with other modern views called by the same name.) Like the followers of Plato, they thought that the categories existed prior to any particular cats or horses. The categories were like original patterns or archetypes. They were the universal patterns that explained why all cats share common features. Each cat, when it came into existence, conformed to the prior pattern of the universal category, which might be called *catness*.

Medieval realism started with the unity of a category. So how did it explain diversity? The medieval philosophers believed in God, so they believed that God creates each cat. He uses the same pattern, namely catness. But if he uses the same pattern, why does each cat not come out exactly the same, like candies made using the same mold (the same pattern)? Even candies made with the same mold show minute differences, which may be due to imperfect mixing of the ingredients, or slight differences in the making process. So a person could try to say that the cats are different because the matter used to make them is different, or the making process shows slight differences. But this explanation just pushes the problem back in time. What generated the differences in the matter? What generated the differences in the processes? The processes presumably have a universal category to describe the unity that belongs to them. So what leads to the differences when we compare two instances of the *same* process?

Opposite to the medieval realists were the *nominalists*. They said that the many was prior to the one. We start out with many cats in the world. Then we give them a common name, the name *cat*. According to the nominalists, the name is nothing but a name. (The word *nominalism* is cognate to the Latin word *nomen*, which means "name"). A name like *cat* does not label a universal category that is out there in the world. The category of *catness* is only in here in our minds. We have invented it. And its invention depends on the prior existence of the many cats out there. Clearly, nominalists think that the diversity of cats is first, and the unity of the category is derived.

Nominalism had the opposite problem from realism. Its problem was to account for the unity. We start with many cats. Why is there anything in common between the many cats, any commonality that would lead us to group them all under a single category of "cat"? Nominalism suggested that the category is our invention, corresponding to nothing out in the world. It is simply an idea. It is an illusion. Or, if a nominalist did not want to go this far, he could say more guardedly that the unity is a secondary construction, based on the primary reality of the diversity of cats. But if we start with pieces that are purely diverse, how can we later create unity? Even if the unity is pure illusion, we need to explain where the unity in the illusion came from. Moreover, it is not plausible to claim that there is nothing "really" similar about the different cats.

Unity and Diversity in the Trinity

According to Trinitarian thinking, the unity and diversity in the world reflect the original unity and diversity in God. First, God is one God. He has a unified plan for the world. The universality of the truth 2 + 2 = 4 reflects this unity. God is also three persons, the Father, the Son, and the Holy Spirit. This diversity in the being of God is then reflected in the diversity in the created world. The many instances to which 2 + 2 = 4 applies express this diversity: four apples, four pencils, four horses, etc. God is the original, while the unity and diversity in the created world are derivative. So we may say that God is the *archetype*, the original pattern, while the instances of unity and diversity in the created world are *ectypes*, derived from and dependent on the archetype.

We can put it in another way. God governs the world by speaking (chapter 1). God has both unity and diversity. So when he speaks—through the Word of God, who is the second person of the Trinity—his speech has unity and diversity. The unities in God's speech specify the unities in the world that he has made; its diversities specify the diversities in the world that he has made.

We can also illustrate unity and diversity in a third way. The unity of God's plan has a close relation to the Father, the first person of the Trinity, who is the origin of this plan. The Son, in becoming incarnate, expresses the particularity of manifestation in time and space. He is, as it were, an instantiation of God. Thus he is analogous in his incarnation to the fact that the universal law 2 + 2 = 4 expresses itself in particular instances like four apples.

What is the role of the Holy Spirit? In addition to other roles, the Holy Spirit expresses by his presence the fellowship between the Father and the Son (John 3:34–35). His role in fellowship has been termed the *associational* aspect.[3] The Holy Spirit is the archetype for the associational aspect. A universal law like 2 + 2 = 4 and the particular instances, like four apples, also enjoy a relation of *association*. The one inheres in the other. In general terms, the associational relation between the one and the many that instantiate the one is an ectypal associational relation.

The Numerical Character of the World

God's plan is the source for the numerical character of the world, as it is the source for every aspect of the world. God's plan is consistent with his character and reflects his character. He is Trinitarian in his character, and so his plan exhibits unity and diversity, and the unity and diversity in the world arise as a result.

In God we find the foundation for numbers. In the world that God has created, we sometimes deal with one, two, three, four, or more apples. Why? Because there are many apples in the world. The apples have diversity. They also have unity. They all belong to one class, the class of apples.

When we have four apples on the table, and we wish to count how many there are, we have already made the decision to treat all the apples on the table as members of one class, the class consisting of the apples on the table. This class has its own unity and diversity. It has the unity of being one class, and the diversity of the four apples in the class. The four apples *belong to* one class, exhibiting the associational aspect. Counting is possible only when we have the unity of one class (the four apples taken together), the diversity of members in the class (each apple), and an associational relation of belonging: that is, the individual apples belong

[3] See the further discussion in Vern S. Poythress, "Reforming Logic and Ontology in the Light of the Trinity: An Application of Van Til's Idea of Analogy," *Westminster Theological Journal* 57/1 (1995): 187–219, reprinted in Poythress, *Logic*, appendix F5; Poythress, *Logic*, chapter 18.

together with the other apples on the table, and they all belong to the same class.

In sum, in our everyday experience, the very idea of number depends on features of the world that embody unity, diversity, and associations. God is the archetype for unity and diversity and association. What we see in the world is the effect of God's word, expressing his plan and his character. He has made the world with ectypal unity and diversity. The combination of these gives us the numerical character of the world.

We have collections of one, two, three, four, or more apples. And we have collections of one or more pears or peaches or pencils. Every class of four apples is an instantiation of the idea of "having four members." The number four expresses the commonality among all instances of four apples, peaches, and the like. In this respect, the number four is the one, showing the unity belonging to all the instances. The instances are the many, showing the diversity. The relation between the unity of the number four and the diversity of four apples or peaches or pencils is an associational relation. Thus the number four depends on the unity and diversity in the Trinity.

The same, of course, is true of any other natural number: one, two, three, four, and so on. Each number, such as 114, is a unity, and the collections of 114 apples or 114 peaches are diverse instantiations of the unity.

Now we can notice another unity in diversity and diversity in unity. All the natural numbers together have a unity. They are all natural numbers! And they have a diversity: each one, such as 114, is distinct from the others.

All of this is so natural, so ordinary, that we are accustomed to taking it for granted. But we can thank God for it. God made it so. Because God is stable, faithful, and consistent with himself, the numbers are stable and the relations of unity and diversity are stable. We live in a world, rather than an absolute chaos. More specifically, God made it so by his word, specifying that it would be so. God speaks. He speaks according to his Trinitarian character. Numbers reflect his character. By reflecting his character, they show us who God is:

> For what can be known about God is plain to them, because God has shown it to them. For his invisible attributes, namely, his eternal power and divine nature, have been clearly perceived, ever since the creation of the world, in the things that have been made. (Rom. 1:19–20)

3

Naturalism

The problem of the one and the many has a long history. But is it still meaningful in our day? Many people would say that it is not. It seems to them like an artificial problem. They might say that it was a question that occupied philosophers before we really understood the nature of the world. But we do not need such things now. We have science to give us answers.

Science gives us fascinating insights and shows its usefulness through the technological spin-offs that we enjoy. But does science give us answers to the most fundamental questions? It does only if we extrapolate science beyond its core achievements.

The Difference between Limited Science and Naturalism

In the Western world of our day, the philosophy of *naturalism* or *materialism*[1] has come to have a wide influence. *Materialism* says that the world consists in matter and energy and motion. It says that there is no God. Or if some kind of god exists, he is irrelevant.

Naturalism or materialism gains prestige from science. Science, it is said, tells us the way things really are. It tells us that the universe all boils

[1] The term *naturalism* is sometimes used more broadly to describe any philosophy that says that the world of "nature" is all that there is. (This view implies that there is no God.) *Materialism* is a particular form of naturalism that says that nature reduces to matter and motion. For simplicity, we are using the two terms as virtual synonyms. To complicate matters further, the word *materialism* is also used to describe a commitment to and fascination with money and material things. This kind of commitment is a serious problem in our time, but is outside the scope of our discussion.

down to matter and energy and motion. But this kind of argument fails to notice a gap between two conceptions of science. In the first conception, science as a discipline confines its attention to matter and energy and motion, in order to study them in depth. Matter and energy and motion—and complex arrangements of them into biological cells and geological formations and stars—become the focus of study.

Then, in a second conception of "science," this focus for science is postulated to be the only thing that really exists. According to this second conception, "science" tells us that the universe is matter and energy and motion and nothing more. But in the process, the word *science* has changed its meaning. People have imported into the meaning a philosophical assumption, namely the assumption that the chosen focus for the practice of science is the *only* legitimate focus, and that it leaves out nothing that is important. This conclusion is not actually the product of detailed investigations into chemistry or star formation. It is an extra hidden assumption. It can never really be justified by detailed scientific experimentation, because such experimentation already presupposes the limited focus on matter. In the nature of the case, it cannot make pronouncements about that which it has not studied.

Science as Focused

Does science with a limited focus answer the problem of the one and the many? No, because the problem of the one and the many is a philosophical problem that is deeper than science. Scientific investigation starts with the assumption that the world is both unified and diverse. Typical experimental science uses the assumption repeatedly. A scientist compares a single experiment on a single bit of matter to other experiments of the same kind on other bits of matter. That is one of the principles about repeating experiments. But to repeat an experiment, the scientist has to rely on the fact that it is identifiable as the *same* experiment. There must be a basic unity. At the same time, the repetition implies that there are two or more instances of the experiment. The multiple instances show diversity. The different instances represent the *many*. The scientist thus is using the interlocking of one and many. He presupposes it rather than explaining it.

Materialism Trying to Answer the Problem

Materialism is the philosophical extension of science that says that there is nothing except matter and energy and motion. Can materialism answer the problem of the one and the many?

On one level, materialism says that human beings evolved in such a way that they see the world as one and many. The one and the many come about as part of our subjective perception. But is the world "out there" actually one and many? Or it is merely that we "see" it that way? Is our way of seeing just an accidental byproduct of mutations and chemistry in our brains? Let us suppose that our way of seeing is an illusion. Then what about the repeated experiments that scientists perform? The idea of a repeated experiment relies on the one and the many and their interlocking. Thus, the idea of a repeated experiment is also an evolutionary illusion, and therefore the science that is built on this way of seeing is an illusion. Since materialism claims to be built on science, materialism itself is also an illusion. This is not good news for materialism.

In fact, most materialists think that the world "out there" is one and many. At the level of material particles, there are many particles. And particles of the same kind share common properties, which means that there is a oneness or unity to all the particles of the same kind.

So where did the unity and diversity come from? Materialists would say that it came from the Big Bang, which through complex quantum mechanical processes led to the creation of a huge number of particles.

There are different kinds of particles. At the level of particles that make up atoms, there are three basic kinds of particles: protons, neutrons, and electrons. They all show some similarities in behavior (for example, they all have a "spin" of 1/2). So there is unity. But the three types differ as well. So there is diversity. There is additional diversity because there are huge numbers of particles of any one of these types.

Or we can go inside the protons and neutrons and say that each proton and each neutron is made up of three quarks. The quarks (in the current state of physical theory) are of six kinds ("flavors"), named whimsically "up," "down," "strange," "charm," "bottom," and "top." In addition, they come in three "colors." We see unity in the fact that all these kinds are quarks, and diversity in the fact that there are different kinds of quarks.

What, then, is the origin of this unity in diversity? Materialists have no complete explanation for the Big Bang itself, the initial event. But they would say that the laws of physics explain the unity and diversity in particles that we see today. Given the Big Bang plus the laws of physics, they would say that we can expect the unity and diversity among the particles. And this unity and diversity in the particles eventually gives rise to unity and diversity at all the other levels, including the levels of ordinary human observation. It sounds good, until we ask more questions.

Where do the laws of physics come from? As human formulations, they contain massive unity and diversity built into them. This unity and diversity comes from the ability of human beings to understand unity and diversity in their minds. But physicists have arrived at the present formulations by interacting with the world, which already had the unity and diversity in its particles. The unity and diversity in the particles leads to the unity and diversity in the formulation of the laws. And then these laws are supposed to explain the unity and diversity in the particles! It looks circular.

The obvious answer is to distinguish human formulations of the laws from the laws themselves—the laws "out there," governing the universe. The human formulations are chronologically subsequent to the existence of the particles. The particles are chronologically subsequent to the existence of the laws "out there." So the laws "out there" explain everything else.

What Explains the Laws?

The laws out there already display the interlocking of unity and diversity. There are at a low level several laws, one for each kind of particle. It is hoped that physics can arrive at a final, unified formulation, sometimes called "the Theory of Everything." Even if it did, the theory would contain unity and diversity within it. It would be one theory, and its oneness would exhibit unity. At the same time, it would be a theory that applied to all the different kinds of particles. The different kinds of particles represent diversity.

In addition, the diversity has to be present in another way. The very conception of a physical "law" implies unity and diversity. Each law is a unity. And each law applies to many particular instances of phenomena in the world. The application represents its diversity.

It should be evident by now that physical explanations do not get rid of the philosophical problem. They just promote the problem to another level. Instead of dealing with the problem on the level of ordinary human experience, we "promote" it into a problem about material particles. Then we promote it from there to a problem about the nature of physical laws.

If we focus on the laws "out there," in distinction from our later human formulations, the laws show the attributes of God, just as the truth about 2 + 2 = 4 showed his attributes.[2] The presence of the interlocking of unity and diversity in the laws reflects the archetypal unity and diversity of God. We have not escaped God by promoting the problem up into the laws.

There is an additional problem. For their formulation, the laws of physics require mathematics—rather advanced mathematics. The advanced mathematics is built up, layer by layer, starting from conceptions of number and space. The investigation of number has already turned up the problem of unity and diversity in its midst. And of course our expression "layer by layer" implies multiple layers, which implies diversity (more than one layer) and unity (a unity where higher layers build on and are in harmony with lower layers). So if we explain the physics by using mathematics, we have promoted the problem of unity and diversity one more stage, into mathematics. We have not "solved" it.

In addition, we confront still another form of unity and diversity. There is unity between the mathematics that the physicists use and the physics to which they apply it. The mathematics "works" when applied to the real world. At the same time, the mathematics is not identical with the physics. Some mathematics has direct physical application, and some does not. Why does there exist a universe to which the mathematics applies?

The naturalist would say that the universe exists because of the Big

[2] Poythress, *Redeeming Science*, chapter 1.

Bang. But we are not really asking about the Big Bang. We are asking, "Why is there such a thing as physical law, as distinct from a purely mathematical truth?" The Bible has a clear answer: God spoke the world into existence. He specified laws that are in harmony with his character. The inner harmony of his nature is reflected in the harmony between mathematics and physics. By contrast, materialism has no answer. In materialism, the laws of physics have to function as a substitute for God.

Materialism is an awkward philosophy. In its typical formulation, it says that nothing exists except matter and energy and motion. But what it really ought to say, at the very least, is that there is matter and energy and motion, and in addition *laws*. The laws are neither matter nor energy nor motion, but something else, an immaterial, conceptual something, involving mathematics.[3]

The Origin of Mathematics

Can we say that mathematics originated because there are multiple objects in the world? We as human beings learn mathematics with the help of our minds, combined with help from an environment that contains multiple objects. But combining our minds with our environment still does not produce an adequate explanation for mathematics. To be sure, we as human beings come to learn mathematics gradually. But the materialist has to have mathematics behind everything else. The mathematics has to be there before the universe existed. And it has to be in harmony with physical laws, that is, laws that are not merely mathematics but actually control what happens, should a universe ever come into existence. To an ordinary observer, this combination of mathematics and physical laws formulated in mathematics is a very complex combination of ideas. It is *ideas*, not matter. It sounds highly immaterial, and highly *nonnatural*, in the sense that the mathematics and the laws of physics are not just "part of" nature. They have to exist in order to call nature into existence in the first place.

In short, mathematics has to be there already, if the materialist is to have any hope of constructing a plausible philosophy. If mathematics in

[3] On materialism, see also Poythress, *Inerrancy and Worldview*, chapter 3 and pages 229–230.

fact testifies to God, as we have argued in the previous chapters, materialism is hopeless.

Some people have hoped to give a more ultimate explanation for mathematics itself by reducing it to logic. We will look at this attempt later. For the moment, we can content ourselves by observing that, even if this attempt could succeed, it would only push the problem of the one and the many back into logic. Logic needs the one and the many to get its foundations started. So this route does not solve the problem.

4

The Nature of Numbers

What is the nature of numbers? And what is the nature of the truths about numbers, truths like $2 + 2 = 4$?

Arithmetic Specified by God's Speech

We have already begun to answer the question by observing in chapter 1 that the truth $2 + 2 = 4$ has the attributes of God, such as omnipresence, eternity, omnipotence, and truthfulness. God spoke the world into existence. As one aspect of his speech, he specified the numerical character of the world. His speech reflects his character. So the truths that he speaks have his attributes.

But we can also see that truths like $2 + 2 = 4$ are accessible to our minds. And we can see that the truths have a bearing on the world around us. Two apples plus two apples is four apples. The truth that $2 + 2 = 4$ is transcendent, in the way that God's speech transcends the world. At the same time, it is immanent. It holds for apples.

Perspectives on Realms

Three different realms come together when we look at $2 + 2 = 4$, namely the realm of transcendent law, the realm consisting in things within the created world (apples), and the realm of our minds. These three are in harmony. To be sure, our minds are not infallible. We can make mistakes in arithmetic. But we can also correct mistakes. And we know that there are formulas that are correct and others that are not. We can come to know that 2 + 2 = 4 and that 2 + 2 does not equal 5. When our minds are in good working order, they match the transcendent law (2 + 2 = 4) and they match realities about apples. Why?

God has ordained all three. He specifies the general truths (2 + 2 = 4). He created the world with apples in it. And he created human beings with minds. The three realms enjoy a fundamental harmony with one another, because God is in harmony with himself and is consistent with himself. He created a world that is consistent, according to his plan.

Perspectives on Ethics

John Frame's three perspectives on ethics can help us to appreciate this harmony. Frame argues that we can approach questions in ethics from at least three different perspectives, with three different starting questions.[1] The *normative* perspective asks what are the norms for ethics. It focuses on God's commandments, such as the Ten Commandments. The *situational* perspective asks what promotes the glory of God within our situation. The *existential* perspective, also called the *personal* perspective, asks what our attitudes and motives should be. It focuses on us as persons.

According to the Bible, these three perspectives are in harmony because God ordains them all. He speaks the norms; he creates the situations; and he creates the persons who are in the situations. Not only are they in harmony, but each points to and affirms the other two. The norms in Scripture tell us to love our neighbors, which is an attitude. So they tell us to pay attention to our attitudes, which are the focus of the existential perspective. And to love our neighbors in action, we have to assess the situation and ask what actions would help them in their situation. So the norms in the Bible push us to pay attention to the situation. Or suppose that we start with the situation. We ourselves are in a sense in the situation, so we have to pay attention to our attitudes. This attention makes us use the existential perspective. In addition, God is the most important person in our situation. When we pay attention to God, we have to pay

[1] John M. Frame, *Perspectives on the Word of God: An Introduction to Christian Ethics* (Eugene, OR: Wipf & Stock, 1999); John M. Frame, *The Doctrine of the Christian Life* (Phillipsburg, NJ: Presbyterian & Reformed, 2008).

attention to his norms, what he desires our moral activity to look like. When we pay attention to his norms, we are using the normative perspective. (See diagram 4.1.)

Diagram 4.1: Frame's Three Perspectives on Ethics

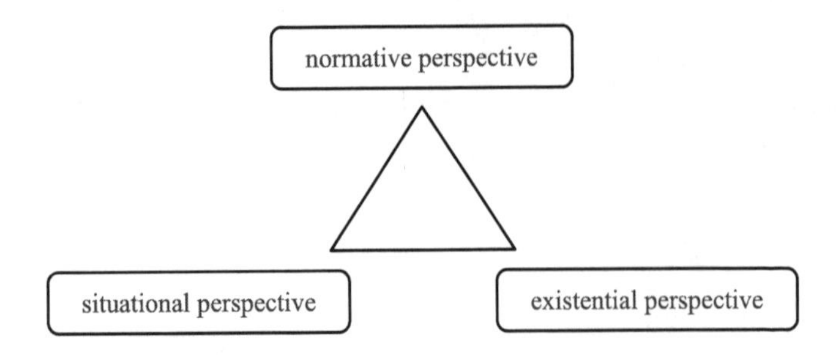

These three perspectives are relevant not only for ethics, narrowly conceived, but for all of life. All of life requires ethical responsibility. When we use these perspectives on the whole world, we find that we are affirming the observations we already made about 2 + 2 = 4. The equation 2 + 2 = 4 represents (1) a norm, (2) a truth about the world, and (3) a truth that we as human beings can know. It involves all three perspectives, normative (law), situational (the world), and existential (us). These three are in harmony.

Not only are they in harmony, but they lead to one another. Each implies the other two. For example, the norm that 2 + 2 = 4 implies that two apples plus two apples will always be four apples, in the world in which we live. Thus the normative perspective implies the situational perspective (which includes apples). In addition, the norm that 2 + 2 = 4 implies that we as knowers should think in conformity with the truth 2 + 2 = 4. The normative perspective implies the existential perspective.

Now let us start with the existential perspective. If we know that 2 + 2 = 4, using the existential perspective, we know that it is true even before we knew it or any other human being was alive to know it. Our knowledge implies transcendence. 2 + 2 = 4 is always true. It is a norm. Thus the existential perspective, which starts with our acts of knowing, leads to the normative perspective, which focuses on the transcendence of the truths that we know. In addition, if we know that 2 + 2 = 4, we can infer that two apples plus two applies is four apples. The existential perspective, which focuses on our knowing, leads to the situational perspective, which focuses on apples.

Perspectives on Numbers

We can use the same three perspectives on a particular number, such as two. It is easiest if we start with the situational perspective. The number two has a relation to all the collections that have two things. These collections embody and illustrate the number two. We can consider two apples, two pencils, two cats. These collections are in the world. We notice them when we use the situational perspective. The situational perspective naturally leads to our seeing the number two as a tool for dealing in a practical way with the collections of things in the world.

Second, let us consider the existential perspective. We as human beings have to be able to think about two for any of these observations about the world to be meaningful to us. We have the word *two*, and we know how to use it. We can observe collections of two objects, and we can think about what it means to say that there are two objects. We have a general conception of the meaning of the word *two*. This conception of *two* within our minds can be distinguished from the collections "out there." Yet it is also related to the collections. We understand the meaning of *two* partly by considering its relations to the collections. In fact, this relationship represents one instance of one and many, unity and diversity. The unity here is the unity of the number two, a unity that exists in its applications to all the collections of two things. The diversity is the diversity of the collections—apples, pencils, cats, and so on.

Third, let us consider the normative perspective on the number *two*. The number two is common to all the collections of two things. It is the *same* number. It represents a pattern of general thinking, and there are *norms* for the pattern. The number two correctly describes some collections, but not others. And there are norms for the use of the number two in relation to other numbers. The laws of arithmetic are norms. These norms include all the arithmetic truths in which the number two appears. The norms have to be distinct from our minds, because in our minds (or in doing arithmetic on paper) we can make mistakes. The norms also have to be distinct from the world, because they are general truths, not just collections of two things in the world.

Other Perspectives on 2 + 2 = 4

We can use other perspectives to consider the richness of the world that God has made. For example, we can consider $2 + 2 = 4$ from the perspective of its relation to other truths internal to mathematics: other arithmetic truths, truths about fractions, about multiplication (for example, $2 \times 2 = 4$), and truths in higher mathematics that rely on the basic truths of arithmetic. All the truths of mathematics cohere in a harmonious way.

Next, let us look at $2 + 2 = 4$ from the perspective of experience in time. We can count objects. Suppose that there are four apples on the table, in two groups of two each. We count one group: one apple, two apples. We count the second group: one apple, two apples. Having finished, we count the whole collection: one, two, three, four. Through this process we have verified that $2 + 2 = 4$. This process shows that $2 + 2 = 4$ has a relation to time and temporal development. In fact, the philosopher Immanuel Kant claimed that human intuitive knowledge of number originated from the perception of time.

We can in fact distinguish two distinct perspectives that focus on time. The first is our subjective experience in time. We do counting in time, and we have a subjective experience of the passage of time, which includes the passage of moments of time or successive heart beats or successive steps in walking. We are aware of the fact that we can count in time. The second perspective is the perspective on the world. The world "out there" has temporal organization.[2]

We can also look at numbers from the perspective of space. The apples on the table occupy different positions in space. From this perspective,

[2] Immanuel Kant maintained that our subjective sense of temporal structure was derived from the categories of our minds, rather than the nature of the world as it is in itself. For a critique of this approach, see Poythress, *Logic*, appendix F1. Time manifests itself both subjectively (the existential perspective) and in the world (the situational perspective).

the truth that $2 + 2 = 4$ is illustrated by considering a single static picture of four apples. They lie there in two spatial groups of two. $2 + 2 = 4$ means that two apples in one spatial group, plus two apples in a second spatial group, together make up a larger spatial group, and this larger group has four apples. As with the perspective focusing on time, the perspective focusing on space can be divided into two, depending on whether we focus on our subjective perception of space (existential focus) or on the organization of the world "out there" (situational focus). In addition, we are aware of a norm: $2 + 2 = 4$ always holds true.

Mathematics in the Physical Sciences

We can also look at numbers from the perspective of physical sciences. The relation of numbers to time and space leads to the use of numbers in the physical sciences: physics, astronomy, chemistry, geology, and biology. Mathematics plays a powerful role in physics, astronomy, and chemistry in particular. Numerical truths are constantly being used. In addition, some kinds of sociology and experimental psychology use numbers and statistics as an integral part of their investigation of regularities in human thought and behavior.

Sets as a Perspective

Next, we can look at $2 + 2 = 4$ from the perspective of sets. We have been talking about collections of apples. Can we generalize the idea of a collection? In mathematics, a *set* is an abstract generalization of our intuitive idea of collections. In effect, we start with a collection, and then in our minds "strip away" all the information except the information about what the members of the collection are. In mathematics, a set whose elements are 1 and 2 is represented by enclosing the list of elements in braces: $\{1, 2\}$. The *members* or *elements* of a set are the items included in the collection. Thus the set $\{1, 2\}$ has members 1 and 2.

Using sets, we can reexpress the truth that $2 + 2 = 4$. As applied to sets, it means that if we have a set with two elements, and a second set with another two elements, which are distinct from the elements in the first set, and then we make another set that contains all the elements from both, this new set will have four elements.

For example, suppose that on the table we have distinct pieces of fruit: one apple, one peach, one banana, and one pear. We group these fruits into two sets. The first set, which we call set *A*, has as its members the apple and the peach: *A* = {apple, peach}. The second set, which we call set *B*, has as its members the banana and the pear. *B* = {banana, pear}. From these two sets we form a third set, *T*, which has as its members all the members of *A* and *B*. *T* = {apple, peach, banana, pear}. This set is called the set *union* of *A* and *B*. In general, the set union *T* of two sets *A* and *B* includes all the members that are in either *A* or *B* or both. In the particular case we are considering, *A* and *B* have no common members. *A* and *B* each have two members. The set union *T* has four. This fact illustrates the principle that 2 + 2 = 4. (See diagram 4.2.)

Diagram 4.2: 2 + 2 = 4, Illustrated by Sets

Are we just saying the same thing that we have already said? In some ways it sounds the same. But we can see that the result, namely the fact that the union *T* has four members, is independent of the details about which kinds of fruit were on the table. We can see that 2 + 2 = 4 is true in general, not just for a particular choice of fruits. We can do general reasoning about sets (ignoring the details about which fruits we are using, or which members we are using that are not fruits, but vegetables or rocks or other objects). The general reasoning about sets shows that the numerical truths have general validity.

In fact, Alfred North Whitehead and Bertrand Russell used this very route in order to try to derive the properties of numbers using logic as the starting point.[3] From logic they proceeded to develop an idea of predicates. (Predicates are abstract representations of properties like "is red" or "is a mammal.") Next, for each predicate they defined a class consisting of the objects that have the property signified by the predicate. Two classes represent the same number if their members can be put into one-to-one correspondence. By this means numbers can be represented through the classes.

Without endorsing Whitehead and Russell's philosophy, we can use logic as a perspective on numbers and a perspective on mathematics as a whole. Much of higher mathematics today is presented using axioms and deductions from axioms. The rigor of formal logic is used to provide a rigorous basis for the mathematics that is presented. Mathematics is indeed logical, and mathematicians can show how vast conclusions can be deduced from simple axioms, appropriately chosen.

Next, we can use language as a perspective on mathematics. Mathematicians communicate using language. They supplement ordinary language with special mathematical symbols, so that mathematical results are a product both of language in general and of the ability to produce new symbols with special mathematical meanings. Even the symbols "+" for "plus" and "=" for "equals," which we use in the formula $2 + 2 = 4$, are special symbols. There are many other mathematical symbols, some of which are less well known because they are used only in specialized areas or only in advanced mathematics. The special symbols are made possible by the flexibility of ordinary language, which allows us to supplement it with newly invented symbols, and allows us to define the meaning of the new symbols. Mathematics can therefore be analyzed as a part of natural language.[4]

Axiomatic mathematics can also be analyzed as a kind of specialized language, a *formal language*, that has specialized rules for deriving conclusions from premises. The perspective of formalized language is

---

[3] Alfred North Whitehead and Bertrand Russell, *Principia Mathematica*, 2nd ed., 3 vols. (Cambridge: Cambridge University Press, 1927).
[4] Poythress, "Tagmemic Analysis of Elementary Algebra."

useful in analysis of mathematics, and is closely related to the perspective of logic.[5]

Mathematics and Social Interaction

We can also consider the relation of mathematics to social interaction. Before they learn arithmetic, children have already developed some basic intuitions about numbers through interaction with the world, with their parents, and with others. They know that there is a difference between having two teddy bears and having one, and they may learn the names for the first few numbers—one, two, and three, and maybe more. Either from parents or at school, they learn more through social interaction. The teachers teach them some arithmetic, and they also interact with fellow students both in the classroom and on the playground, where they may sometimes play games that use numbers within the games.

Social interaction also takes place among professional mathematicians. A mathematician may interact either with fellow mathematicians or with scientists in physics, chemistry, computer science, or other sciences in looking for suitable problems to solve, and in exploring how best to solve them. If he develops a new approach to a mathematical problem, or offers a proof of a new theorem, he interacts with mathematicians by presenting the approach or the proof for their inspection. Sometimes other mathematicians find a gap or a problem in a proof, or even a counterexample: the proposed proof fails. Mathematicians show their human limitations in the fact that a single mathematician working by himself may not see the gap that others find later. Mathematics depends on social interaction for verifying the work of individual mathematicians.

Teachers of mathematics must also pay attention to pedagogical issues. How can they present concepts and methods in mathematics so that they make sense to new students, and how can they give the students not only facts and rules but insights and interest? How can a teacher bring discipline and order into an unruly classroom, so that the students are in an environment where they can concentrate on learning? How can teachers motivate students who do not care whether they pass, and do

[5] Poythress, *Logic*, chapters 55–57.

not see the point? People are complicated, and skill in teaching involves much more than competence in knowing the subject-matter, in this case competence in some area of mathematics.[6]

Cultural Influence

We may also consider the influence of mathematics on larger social and cultural issues. Some people consider mathematics a key example of rigorous, clean thinking. So it becomes a model for how science should proceed. Or it becomes a model for science in a second sense, that a piece of science that uses numerical calculations has more prestige and receives more attention and admiration than a piece that does not use numbers. This effect can be observed in social sciences, where it biases some practitioners to prefer to study only measurable or quantifiable aspects of human interaction. The result may be that only what is measurable counts as scientifically "significant." And then, if science is also the model for all knowledge, only what is measurable counts as significant for all of life. This kind of tendency in thought does not reduce everything to matter and motion, the way that materialist philosophy does, but rather it reduces everything to quantity and measurement.

We may also ask what may be the influence of a secularist conception of mathematics. If our culture conceives of mathematics as existing out there independent of God, and it thinks that God is irrelevant to mathematics, does this assumption tend to reinforce secularizing forces all across society? If God is irrelevant to mathematics, then maybe he can be made irrelevant to all other sectors of society, if we succeed in analyzing them mathematically.

Harmony between Perspectives

God has ordained that numbers function in relation to all the perspectives that we have considered—and more as well. He has ordained the truth that $2 + 2 = 4$ as a permanent, universal truth. He has also established it in

[6] Helpful material on the social, pedagogical, and cultural aspects of mathematics can be found in Russell W. Howell and W. James Bradley, eds., *Mathematics in a Postmodern Age: A Christian Perspective* (Grand Rapids, MI/Cambridge: Eerdmans, 2001).

relation to many other truths: truths of arithmetic, truths in higher mathematics, truths in physical sciences, truths about collections of apples, truths about people in their social interaction and pedagogy, and truths about cultural influence (see diagram 4.3). These all enjoy harmony with one another because they originate from one God who is in harmony with himself.

In a philosophical approach informed by the God of the Bible, we can enjoy the richness of the world. The world has many dimensions, many complexities, and many beauties. We have no need to try to explain the richness of the world by deriving it all from one aspect. We do not need to say that arithmetic generates everything else in the world. Nor do we need to say that logic generates everything. Or physical objects. Everything is what it is. Everything is unique. And everything is related to everything else. God's plan and God's rule over everything produces both coherence and distinctiveness. The coherence and distinctiveness represent another expression of the one (coherence) and the many (distinctiveness).

Nothing is *reducible* to anything else. Our approach opposes *reductionism*, the philosophical attempt to claim that one aspect of the world is the most ultimate and that everything ought to be explained completely from this one aspect. For further discussion of the general principle that the world is rich and that reductionisms are inadequate, we must direct people to the fuller discussion of what philosophy and metaphysics look like when they are reformed by the Bible's instruction.[7]

For example, naturalism or materialism, which we discussed earlier (chapter 3), tries to reduce everything to the *material* or physical level. It claims that everything is "really" matter and energy and motion. Another philosophy, called *empiricism*, tries to reduce everything to sense experience. Still another philosophy, *idealism*, tries to reduce everything to ideas in the mind.

All of these reductionistic philosophies have difficulties. The most basic difficulty is that things are different from one another. Even though there is impressive harmony, nothing is really explained in all its

[7] Poythress, *Redeeming Philosophy*.

Diagram 4.3: Multiple Relationships

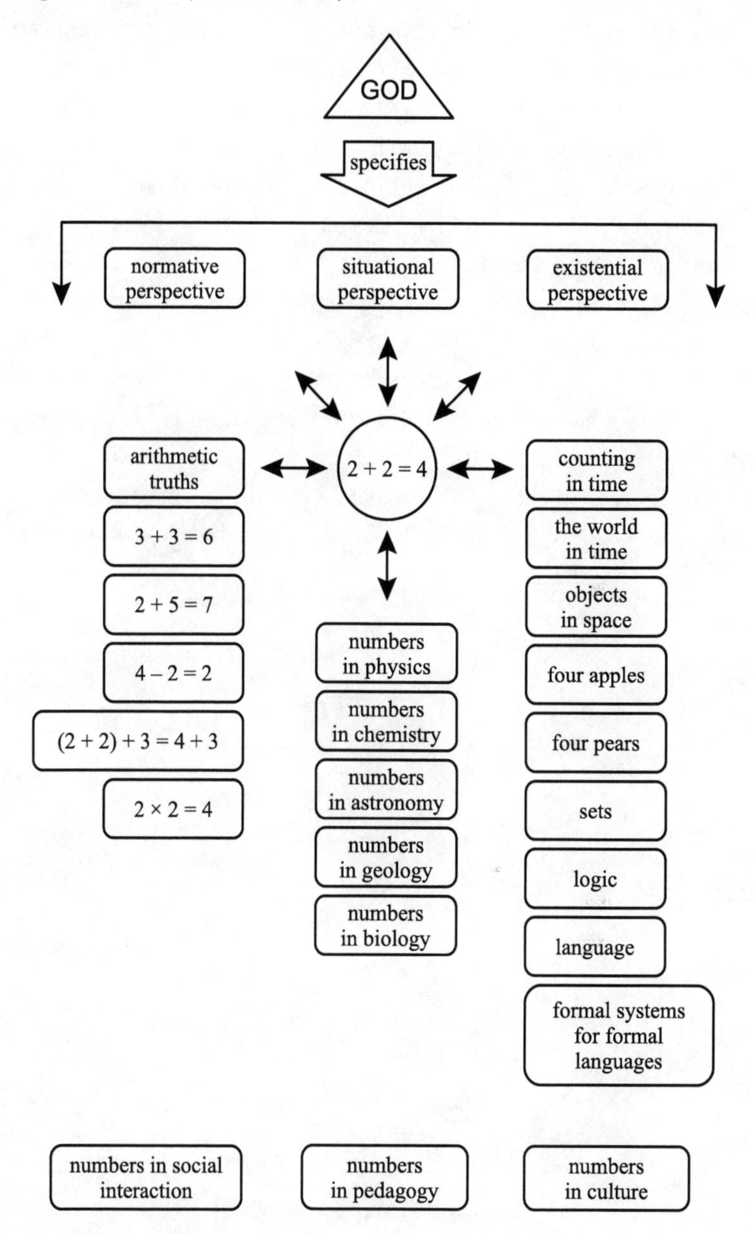

dimensions through a reductionism. The materialist claims that a rainbow is nothing but physical light rays acting according to physical laws of refraction. But has he explained the beauty of the rainbow? The empiricist says that the rainbow is nothing but visual sensations conveyed to the brain from the retina and optic nerves. Has he explained the beauty? The idealist says that everything is in our mind. But has he explained the ways in which the world around us surprises us?

Similar principles hold when it comes to explaining mathematics. Philosophers of mathematics have attempted to explain the nature of number and the nature of arithmetic truth. But most of these attempts have been reductionistic (see appendix A). It is better to appreciate the world as God made it. Numbers and arithmetic truths exist in relationship to a whole world, and this world is multidimensional. God made it that way. There is no reason to fight against it, trying to imagine numbers as they might be if they could be perfectly isolated from the world.[8]

---

[8] The analogous attempt to isolate logical truth from the world is discussed in Poythress, *Logic*.

Part II

Our Knowledge of Mathematics

5

Human Capabilities

If we are to travel further in understanding the ways in which God is the source for mathematics, we need to consider briefly the nature of our capabilities as human beings. What can we hope to understand about God? And how?

The Bible indicates that God is Creator and we are creatures. He is infinite and we are finite. So we cannot understand him exhaustively. The word *comprehend* is used in a technical sense in theology to express the limits of human understanding. Theologians say that we cannot *comprehend* God; that is, we cannot understand him completely, as he understands himself. We can nevertheless know God—in fact, everyone does know God, according to Romans 1:19–21, even those who are in rebellion against him and are trying to suppress the knowledge.

The Image of God

The Bible indicates that God made man in his image (Gen. 1:26–27). We are not merely products of a gradualistic, impersonalistic, purposeless evolutionary process. Being made in the image of God implies that we are like him. In Genesis 1 the Bible does not go into detail about all the ways that we are like God. But from the rest of the Bible we can see that there are many likenesses. We can reason; we have a sense of morality; we can use language; we can make personal commitments; and so on. Alongside many other capabilities, as human beings we are capable of thinking God's thoughts after him. In particular, we can know that 2 + 2 = 4, a truth that is in God's mind before it is in ours.

Because of the distinction between Creator and creature, we can say more specifically that we think God's thoughts after him *analogically*. Our thinking processes are not simply identical with God's. And they do not need to be. God knows everything because he knows himself and his plans. We need to grow in knowledge. We need to observe things in the world, and receive instruction from other people. Through fellowship with God, and through fellowship with people whom God puts in our path, God teaches us knowledge (Ps. 94:10; compare Job 32:8).

We need to affirm both the similarities and the differences between God's knowledge and ours. Let us use the example of 2 + 2 = 4 to do it. God is the origin of the truth that 2 + 2 = 4. It is true because he speaks it. We are receptive of its truth. God knows its meaning exhaustively, in relation not only to its embodiments in four apples and four peaches, but in relation to every other truth whatsoever. We know only some of these relationships, and we know them partially. But we also know truly. 2 + 2 = 4 is indeed totally true, both for us and for God. It is true for us because first of all it is true for God.

Transcendence and Immanence

We can go further in understanding human knowledge by using insights from John Frame. John Frame produced a diagram, now called *Frame's square*, for summarizing God's transcendence and immanence.[1] It expresses the difference between a Christian view of transcendence and immanence, on the one hand, and a non-Christian view on the other. (See diagram 5.1.)

The left-hand corners of the square, labeled 1 and 2, represent the Christian understanding of transcendence (corner 1) and immanence (corner 2), as taught in the Bible.[2] God's transcendence means that he has absolute authority, and that he controls the world. God's immanence means that he is present in the world.

[1] John M. Frame, *The Doctrine of the Knowledge of God* (Phillipsburg, NJ: Presbyterian & Reformed, 1987), 14.
[2] It is important to understand that many people today who would claim to be Christian are confused and do not consistently think and live according to a Christian view of transcendence and immanence. In fact, "Christian" teachers who represent modernist forms of Christianity may teach in accord with the right-hand side of the square, the non-Christian view.

Diagram 5.1: Frame's Square on Transcendence and Immanence

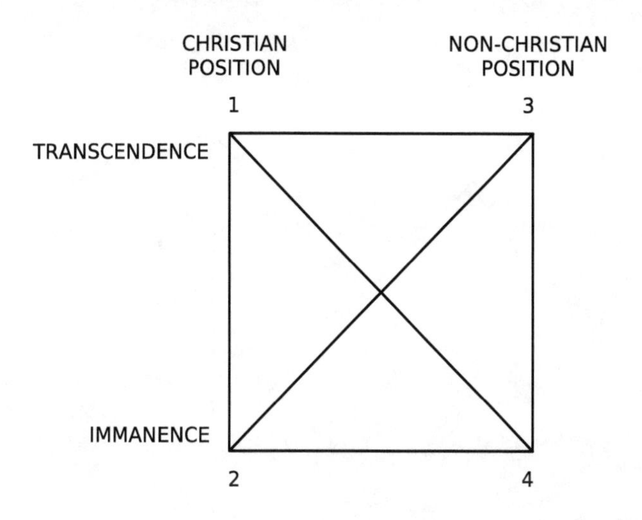

The two right-hand corners of the square, labeled 3 and 4, represent the non-Christian understanding of transcendence (corner 3) and immanence (corner 4). Non-Christians differ among themselves. But much non-Christian thinking maintains that God's transcendence means he is uninvolved (or even that he does not exist). He is remote. A non-Christian understanding of immanence says that God is identical with the world or limited by the world. The two horizontal sides of the square represent the fact that there are superficial similarities between the two sides. They can *sound* the same. Each can use the words *transcendence* and *immanence*. Each side might sometimes say that God is "exalted" (transcendence) or that he is "nearby" (immanence). But they *mean* different things, even when the language is similar. (See diagram 5.2.)

The diagonals of the square represent contradictions. The non-Christian view of transcendence (corner 3), by saying that God is uninvolved, contradicts the Christian view on immanence (corner 2), which says that he is present and involved. The non-Christian view of immanence (corner 4), by saying that God is subject to the limitations of the world, contradicts the Christian view of transcendence (corner 1), which says that he sovereignly controls the world and is not limited by it.

Diagram 5.2: Frame's Square with Explanations

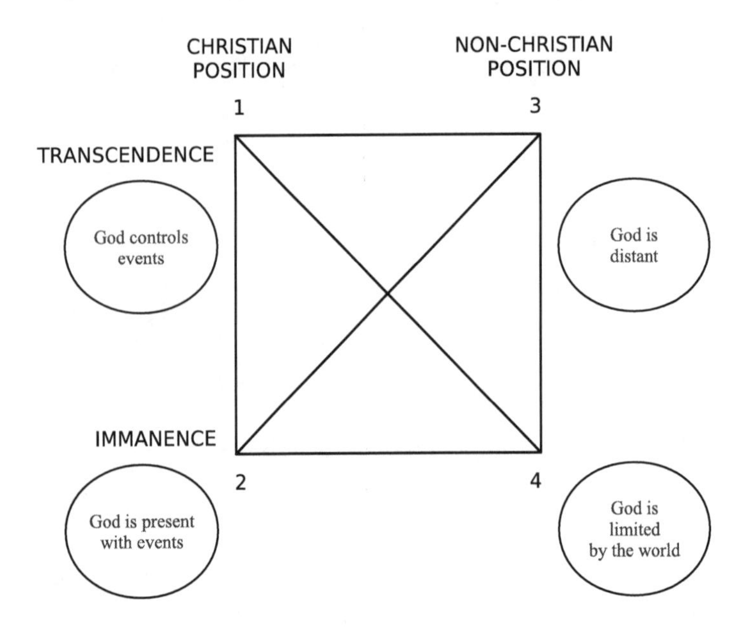

We can apply Frame's square to the understanding of a particular numerical truth, such as 2 + 2 = 4. According to a Christian view of transcendence (corner 1), God's authority stands behind the truth of 2 + 2 = 4. God controls numbers rather than being subject to them. Second, according to a Christian view of immanence (corner 2), God through his presence in the world holds the world to its conformity with the truth 2 + 2 = 4. Two apples plus two apples make four apples because God is present with the apples, expressing his truth. God also is present in our minds, so that we can come to know that 2 + 2 = 4.

Third, according to a non-Christian view of transcendence (corner 3), God is uninvolved with numerical truth—numerical truth is just an abstraction, just "out there" (or maybe just "in here," if truth is completely subjectivized). God is also uninvolved with apples. This non-Christian view contradicts the Christian view of immanence (corner 2). Fourth, according to a non-Christian view of immanence (corner 4), God is limited by numbers. They control him by restricting what he can do. He has no authority over the truth that 2 + 2 = 4. This view contradicts the Christian view of transcendence (corner 1). (See diagram 5.3.)

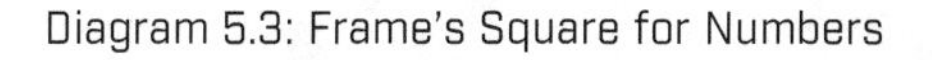
Diagram 5.3: Frame's Square for Numbers

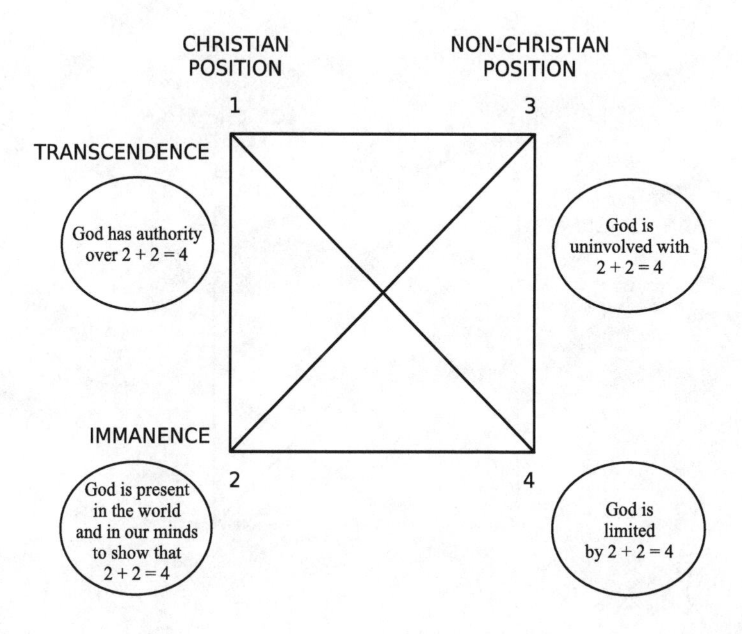

Transcendence and Immanence in Issues of Knowledge

We can apply the same principles when we consider the issue of knowledge. God knows that 2 + 2 = 4. And human beings know that 2 + 2 = 4. What is the relationship?

According to a Christian view of transcendence (corner 1), God is the original, authoritative source and knower of 2 + 2 = 4. He knows it exhaustively. According to a Christian view of immanence (corner 2), God through his presence makes the truth 2 + 2 = 4 accessible to and known to human beings, who know it derivatively and analogically.

According to a non-Christian view of transcendence (corner 3), God does not exist or does not know anything or is uninvolved in human knowing of 2 + 2 = 4. According to a non-Christian view of immanence (corner 4), we as human beings can be the standard for knowing. Our knowledge of 2 + 2 = 4, according to our own autonomous standards, can be used to specify what God's relation must be to the truth that 2 + 2 = 4. As usual, the non-Christian view contradicts the Christian view. (See diagram 5.4.)

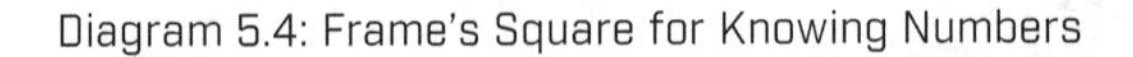
Diagram 5.4: Frame's Square for Knowing Numbers

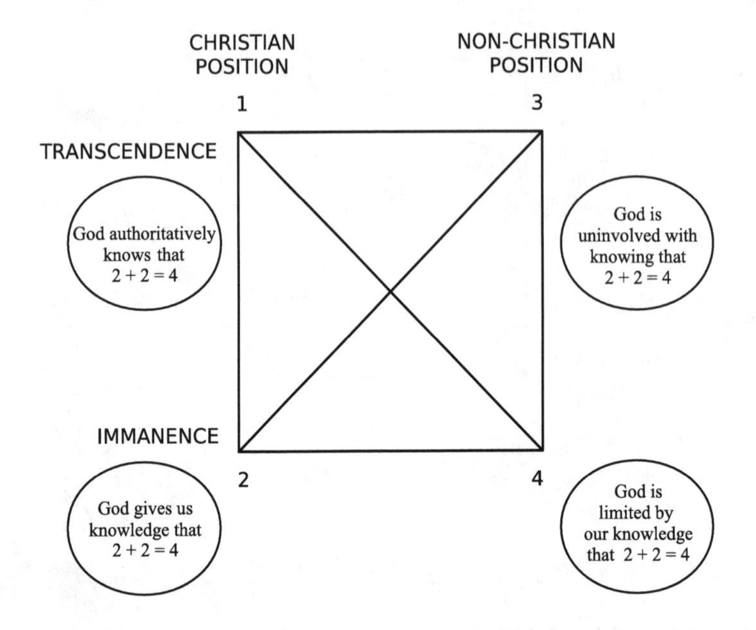

Much more could be said about issues of human knowledge and the human process of coming to know. We must leave that to other books.[3] It is enough for the present for us to understand that human knowledge of numerical truths, such as 2 + 2 = 4, derives from prior, archetypal divine knowledge. It is imitative and analogical. God calls on us to praise him for his infinite knowledge, and to acknowledge his authority in knowledge. This principle includes our knowledge that 2 + 2 = 4. We are to conduct our lives and our thinking about mathematics in the light of God's greatness and his being worthy of praise and glory.

---

[3] See Frame, *Doctrine of the Knowledge of God*; Poythress, *Redeeming Philosophy*; Esther Lightcap Meek, *Longing to Know* (Grand Rapids, MI: Brazos, 2003).

6

Necessity and Contingency

Now we focus on the question of whether mathematical truths are necessary. Is it necessary that $2 + 2 = 4$? Or is this truth something characteristic only of the way that God created our universe? We may ask similar questions about other truths of mathematics, both truths about numbers and truths in more advanced mathematics, such as calculus and group theory.

An Intuition of Necessity

As a first intuitive reaction, many people might say that mathematical truths are necessary. We could not imagine a world in which $2 + 2 = 4$ was not true. Mathematical truths seem to be "basic" and not specific to our universe. By contrast, the existence of apples or the existence of physical laws like Newton's laws of motion has a connection to the particular world in which we live. We could imagine a world where things were vastly different.

Let us suppose that this intuition about necessity is correct, and that mathematical truths are all necessary. Then how does this necessity relate to God?

One route that people have tried is to conceive of this necessity as something independent of God or even superior to God. Allegedly, God is limited by the fact that he must conform to the truths of mathematics, and that any world that he creates would have to conform to the truths of mathematics.

But the idea of God being limited by mathematics makes mathematics superior to God, and in that respect it becomes a kind of "god above God." Its authority is more ultimate than God. If we follow the Bible's teaching about God, that will not do. God is the only Lord, the ultimate Lord. How then could mathematical truths be an *additional* necessity?

Necessity and Contingency with Respect to God

We can find an answer if we reflect on the character of God. The Bible indicates that God is *omnipotent*, all powerful. Sometimes people think that omnipotence means that God could do anything at all, even something contradictory or something morally evil. But that is not right. God cannot do anything that would violate his own character. For example, he cannot lie:

> God is not man, that he should *lie*,
> or a son of man, that he should change his mind. (Num. 23:19)

> . . . God, who *never lies*, promised before the ages began . . . (Titus 1:2)

He cannot deny himself (2 Tim. 2:13); he cannot contradict himself; he cannot do anything morally evil, because that would be inconsistent with his goodness.

Omnipotence, then, does not mean that God could do anything that we could imagine, but that he can do anything that he *wants* to do: "Our God is in the heavens; he does all that he *pleases*" (Ps. 115:3). Since what pleases God or what God wants is always consistent with his character, there is no difficulty.

In particular, since God is himself rational and is the source for human reason, he never does anything irrational. He is never inconsistent. He is consistent with himself, and his self-consistency is the foundation for logic.[1]

In the same way, we can infer that God is the foundation for mathematics. He is the source both for that which is necessary and for whatever in mathematics is contingent.

[1] Poythress, *Logic*, especially chapter 13.

What is *contingent* is whatever could have been otherwise in a different universe, or could have been otherwise if God had never created a universe at all but had simply remained himself. The Bible indicates that God is self-sufficient. He does not *need* the universe. It was not *necessary* for him to create anything outside himself. And, having decided to create a universe, it was up to him to decide what kind of a universe he would create. So the details of this universe—the fact that it has apples and horses and not unicorns—are a product of his free decision. They could have been otherwise.

The Bible indicates that God had a plan even before he began to create, and that his plan for the universe and for history was comprehensive:

> . . . having been predestined according to the *purpose* of him who works all things according to the *counsel of his will*, . . . (Eph. 1:11)

> . . . even as he chose us in him *before the foundation of the world*, that we should be holy and blameless before him. (v. 4)

Given that God has planned something, it is *necessary* that it take place. But this subordinate "necessity" derives from two kinds of sources: (1) God's faithfulness, which is an aspect of his eternal character, and (2) God's plan, which could have been otherwise. So this kind of "necessity" is not on the same level as the necessities of God's character. God's character could not be different from what it is. But his plan could have been different, had he so chosen. Theologians have long distinguished between *necessary knowledge* that God has of his character and *free knowledge* that he has concerning his plan and concerning contingent facts about the world. In both cases God is the absolute God.

Necessity and Contingency with Respect to Mathematics

What are the implications for the truths of mathematics? Some truths of mathematics, or perhaps all truths of mathematics, may be necessary in an absolute sense, because they are implications of God's character and his self-consistency. But we must also consider whether some or all truths of mathematics might be contingent, in that they are a product of the plan that he freely chose for creation and for the development of the world that he created.

As we observed, most people's intuition tends to put mathematical truths on the side of necessity. But we have to be careful, because our minds are not the lords of the world. God is the Lord. Our minds are derivative. God has made our minds so that we are naturally in tune with the world that he has made. But he could have made a very different world from ours. Is it possible, then, that mathematical truths seem to us to be necessary only because we are adapted to a world in which the truths hold? Is it possible that God in his plan specified mathematical truths as *contingent* truths about *our* world? Might some other world have different truths? And what if God had not created any world at all? Would mathematical truths still hold? Or did God bring such truths into existence, as it were, only when he planned to create a world?

Knowing God

Such questions are interesting. But we must be cautious in trying to answer them. We need to recognize our limitations as creatures. Frame's square concerning transcendence and immanence is relevant. On the one hand, when we use a Christian view of transcendence, we cannot presume to dictate to God what is and is not possible for him. We as creatures cannot see into the depths of God to know the exact implications of his self-consistency. We cannot just content ourselves with a surface level of reasoning, in which we say, "Well, mathematical truths seem to *me* to be necessary, so they *must* be necessary for God as well." That kind of reasoning is in danger from corner 4. Corner 4 says that we use our own minds as the ultimate standard for what can be the case, rather than acknowledging that we must receptively submit to God's superiority (corner 1).

According to the Christian view of immanence, God has made himself known. We can be confident on the basis of what he has revealed to us. So we can reason about the status of mathematical truth. We can observe that God has revealed to us his Trinitarian character. He is one God in three persons, the Father, the Son, and the Holy Spirit. When we say, "one God," we use the number one; when we say, "three persons," we use the number three. We talk about the truth in language that God has given us. This language enables us to talk both about God and about the world. God has designed it so that it is adapted to the world.

So people might be tempted to argue that the words *one* and *three* and their meanings belong *exclusively* to this world, and do not really apply to God. "God," they would argue, "is inaccessible in language. God is not *really* either one or three." But now the people who talk in this way have made a transition to corner 3 in Frame's square. They are saying that God is unknowable. Their position does not respect the fact that God has revealed himself. He reveals who he actually is, not a fake or a substitute. Otherwise, we would all be idolaters, because we would be worshiping the fake rather than the true God. If we return to corner 2, the Christian view of immanence, we have to say that God has told us that he is one God in three persons. Both his being one God and his being three persons are eternally true. They did not "become" true merely because God decided to create a world and decided to reveal himself to creatures in the world. We really do know God. He really is one God. There really are three persons, and these three exist eternally. This truth is mysterious to us, because we can never comprehend or know exhaustively the meaning of the Trinity. But we believe that what God says is true. And we are right in doing so. These things are true of God eternally, even though we as creatures have come to know them only in the course of time and in the course of our experience of fellowship with God.

We need to struggle to keep on the left-hand side of the square. It is not always easy to discern when we have begun to drift toward the right-hand side or when we have compromised with the right-hand side. It becomes challenging in particular when we ask deep questions about what is necessary and what is not.[2]

We do not know God *in the same way* that God knows himself. God's knowledge is infinitely deep. God knows his own unity as one God in a way different from the way that we do. He knows his own Trinitarian character and the distinctiveness of each person in the Trinity

[2] On necessity, see also Poythress, *Logic*, chapters 65–66.

in his own unique way as God. The differences follow from God's transcendence.

At the same time, God is one God. The oneness of God exists before there ever was a world of creatures that could understand for themselves the idea of oneness. Likewise, God is three persons. The threeness of the three persons exists before there ever was a world of creatures.

As we observed (chapter 2), God's nature provides the original instance of the one and the many. He is the *archetype*. He is the original pattern. When he creates a world, he creates according to his character and by the power of his speech, his Word. Oneness and threeness and all the other numerical properties that we find in the world are the product of his speech. But that does not imply that every aspect of the numerical properties that we see is *merely contingent*. The numerical properties have contingent aspects, in the sense that they are properties that apply to instances in the world, like four apples and four peaches. But they are at the same time expressions of the faithful and wise character of God. They also express his self-consistency. God's faithfulness and wisdom and consistency are necessary, because they are aspects of his character.

It is not necessary that the truth $2 + 2 = 4$ should be instantiated in a particular case with four particular apples. But it *is* necessary that, if there are four distinct apples, they would conform to the self-consistency of God.

It seems to me that the numbers one and three have a unique role in God himself. We can also say that there are two other persons besides God the Father, so we have a manifestation of the number two as well. We do not have the same for the number four or for larger numbers. God does not need more than three persons in order to be himself.

Can we go further than this point? As usual, we must be cautious. But can we say that God knows all the possibilities for worlds that he could create? Would he know all these possibilities, even if he had decided never to create a world, but just to remain himself? Cautiously, with the voice of a creature, I say that I think so. If so, would his knowledge include the knowledge of numbers of creatures in any world that he might decide to create? I think so.

If, cautiously, we include in our reckoning God's knowledge of possibilities, it seems that we can conclude that numbers exist eternally. Now the Greek philosopher Plato thought that abstract concepts like the concept of the good or the beautiful existed eternally, independent of God. By analogy, an imitator of Plato could postulate that numbers as abstract concepts exist eternally as abstractions, independent of God. But our view of God's absoluteness leads to another view: numbers exist, not as Platonic abstractions, but as an aspect of God's knowledge. And, because of the principle of the one and the many, we can say that they do not exist in God's mind as pure unities, utterly detached from a plurality of possible instantiations. Rather, they enjoy a relationship to the plurality of creatures in worlds that God might choose to create.

But we can still ask whether numbers have to be the same with respect to all the worlds that God might choose to create. The number four has an instantiation in four apples within this world. It would not have the same instantiation in a world in which apples did not exist. The one and the many interlock. Likewise the single number four interlocks with its possible instantiations in this world and in other possible worlds. So there is complexity as well as unity.

Granted this complexity in unity, we could go on to ask whether the laws of arithmetic, such as $2 + 2 = 4$, might actually be different in another universe. Could God create a universe in which $2 + 2 = 5$? My own intuition suggests not. But of course I am a creature; my intuition is not infallible. If $2 + 2 = 4$ is true in any universe that God might create, does it restrict God's omnipotence? Let us remember the meaning of omnipotence. God cannot do anything that would be inconsistent with his character. The basic question is whether the consistency of God's character implies that $2 + 2 = 4$ is true in any universe that he might choose to create. We will take up this issue only after thinking more about the meaning of numbers in *our* universe.

Does It Matter?

Does it matter whether numbers are eternal, or whether they are necessary or contingent? Our reflections in this chapter still leave us with mysteries. As creatures, we cannot comprehend God's Trinitarian character.

Let us not overestimate our capabilities. And let us not suppose that we *need* to know more than God has given us to know and enables us to know. For most purposes, it is enough to understand that all truth comes from God, including the truths about numbers, and that numbers themselves are among the gifts that God has given. They are not Platonically independent of God.

In addition, we need to make sure that we preserve our understanding of the reality of the Trinitarian character of God. God is one God in three persons. The words *one* and *three* have meaning when we talk about God. We should not think that we have to travel beyond numerical meanings in order correctly to describe God. We should not lapse into a kind of thinking where we treat God as distant and unknown (corner 3 of Frame's square). On the other hand, we should also maintain that God is not one and three in exactly the same way as one apple and three apples are. God is God and is unique. Nothing in creation gives us an exact model. These observations maintain the truth of God's transcendence (corner 1 of Frame's square). If we thought that we had an exact model, we would be using the model to try to make God conform to our way of thinking as a standard. We would be following the pattern of non-Christian immanence in corner 4.

If we understand these truths, we need not think that remaining mysteries are a threat to our ability to serve God.

Part III

Simple Mathematical Structures

7

Addition

We can now begin to consider briefly some of the specific areas that are part of mathematics. We want to grow in glorifying and praising God in these areas. One of the areas is arithmetic. Children learn how to add, subtract, multiply, and divide. They start with whole numbers. Later they learn how to add and subtract with fractions and decimals. What does God have to do with learning addition?

Children's Learning in Relationships

As we indicated, children learn through interacting with the world and through interacting socially with other people, especially teachers. So there is a complex social dimension to their knowledge, and this includes their knowledge of addition.

Some children may learn addition by rote. They memorize the addition table. Then they practice applying what they have learned to problems:

> Teacher: 2 + 2 = ? Child: 4.
> Teacher: 2 + 1 = ? Child: 3.
> Teacher: 3 + 4 = ? Child: 6. Teacher: no, 7. Child: 3 + 4 = 7.

But a child who learns just by rote may know nothing about the *meaning* of 2 or 3 or the relationships of numbers and addition to the larger world. In that case, the child will not see how to apply what he learns to practical cases, such as buying apples at the grocery store. To learn the addition table in isolation from everything else is poor pedagogy. And it makes learning harder for the child, because he does not see what is the point.

So children should be learning in the context of life. If so, they are learning in the context of God's world. They are continually relying on the coherence of the world, which God has established. They are seeing richness that God has planned in his wisdom and given to them in his bounty. As we observed in chapter 4, truths about numbers have multiple relationships with many areas of study. Children learn at least a bit about this multitude of relationships, and they absorb a good deal without being explicitly told. God has ordained all these relationships. His wisdom and his bounty are expressed in the numbers. And they are expressed in the truths concerning numbers. The truth that $2 + 2 = 4$, and every one of the truths that the child learns, are truths from God. Every truth reveals the omnipresence, eternity, immutability, omnipotence, and beauty of God, as we have observed (chapter 1).

Children are also learning in the context of the world. The teacher shows them instances of two apples plus two apples making four apples. Children absorb the truth by using the relationships between the one truth, $2 + 2 = 4$, and the many instances (with apples).

Children learn in the context of their own subjective experience. It is they who have the experience of "seeing the point" or maybe of continually struggling when they have not seen it. We can see that the normative, situational, and existential perspectives are pertinent. The child learns normative truths ($2 + 2 = 4$) in the context of illustrations in the situation (apples) and in the context of his own subjective experience (illustrating the existential perspective).

All of us who learned arithmetic learned this way. But elementary arithmetic has become "second nature" to most of us. We have probably forgotten the details of how we learned it.

The easiest way to understand addition is to focus first of all on objects in the world, and the groupings of those objects. The teacher shows the child two groups of two apples each, or two groups of two pencils each.

We know from the Bible that God has created the world so that there are distinct apples and distinct pencils, and that we can group them together. We know that his word specifies all the truths about arithmetic and the relationships of these truths to the many aspects of the world. But can we say more?

Re-creation

There are many perspectives that we could use to deepen our understanding. I focus first on re-creation.[1] In the beginning God created the world, as described in Genesis 1. Adam failed in his task. Christ came as the Last Adam (1 Cor. 15:45). By his death and resurrection, he redeemed a new humanity. Those who trust in him are saved from the corruption and death that Adam brought into the world. They look forward to the bodily resurrection from the dead and to entrance into the new heavens and the new earth that God will create after Christ returns (Rev. 21:1).

Tabernacle and Temple

When Christ became incarnate, "the Word became flesh and dwelt among us, and we have seen his glory, glory as of the only Son from the Father, full of grace and truth" (John 1:14). The expressions for dwelling and glory point back to the tabernacle of Moses (Exodus 25–40) and the temple of Solomon (1 Kings 5–8). These two structures were symbolic dwelling places for God that anticipated the final dwelling of God with man that took place in Christ. John 2:21 confirms the relationship between the temple and Christ by saying, "But he was speaking about the *temple* of *his* [Christ's] *body*."

Christ's coming inaugurated a redemptive re-creation. He healed the blind and the lame, in anticipation of the final healing of the body that will be accomplished in the new heavens and the new earth and the new resurrection bodies that believers will receive in the future, when Christ returns (1 Cor. 15:44–49).

The Old Testament tabernacle and Solomon's temple prefigure these realities. They point forward to Christ as the temple. But they also anticipate the "temple" character of the heavenly Jerusalem in Revelation 21:

[1] I discuss this theme and related themes concerning the tabernacle in Poythress, *Redeeming Science*, chapters 11, 12, 17, and 20. On implications for mathematics, see ibid., chapter 22.

"I saw no temple in the city [Jerusalem], for its temple is the Lord God the Almighty and the Lamb" (v. 22). The tabernacle of Moses and the temple of Solomon accordingly have symbolism that has affinities with the creation as a whole, and in particular with heaven as the dwelling place of God. The lampstand in the tabernacle corresponds to the lights of heaven. The bread on the table for the bread of presence corresponds to the manna that comes from heaven. The ark corresponds to the throne of God in heaven, and the cherubim on the ark corresponds to the cherubim who surround God's throne in heaven.[2]

In sum, the tabernacle and the temple reflect God's presence in heaven. God instructs Moses, "Exactly as I show you concerning the pattern of the tabernacle, and of all its furniture, so you shall make it" (Ex. 25:9). Moses receives the pattern on Mount Sinai, where God comes down from heaven. It is a heavenly pattern. And it is explicitly a pattern to make "a sanctuary, that I may *dwell* in their midst" (v. 8). Solomon in 1 Kings makes the temple as a symbolic dwelling place for God, but in his prayer of dedication he recognizes that heaven is God's dwelling in a more ultimate sense: "And listen in *heaven your dwelling place*, and when you hear, forgive" (1 Kings 8:30). "But will God indeed dwell on the earth? Behold, heaven and the highest heaven cannot contain you; how much less this house that I have built!" (v. 27).

These structures have significance for mathematics because they display simple mathematical relationships.[3] In the tabernacle, the Most Holy Place has the shape of a perfect cube, 10 × 10 × 10 cubits. The Holy Place is 10 × 10 × 20 cubits. Some of the furniture also has simple, harmonious proportionalities in its dimensions.

The simple proportionalities belong to the small house, which is an image of heaven and in fact of the whole universe as the large house filled by God's presence (Jer. 23:24; compare 1 Kings 8:27). The fact that the small house is a copy or image of the big house suggests that the big house may also display harmonious proportionalities. And indeed this turns out to be true, as the mathematical character of basic physical laws attests.

All of these structures derive from God. Their beauty reflects the

[2] See further Poythress, *Shadow of Christ in the Law of Moses*, chapters 1–5, 8.
[3] See further Poythress, *Redeeming Science*, chapter 20.

beauty of God. Their harmonies reflect the harmony of God. It is true of the tabernacle, and it is true of the universe as the large-scale house. The pictorial symbolism in the tabernacle confirms what we have inferred from the explicit teaching of the Bible, namely that numbers in our minds and numbers in the world reflect the numerical ordering that God has normatively specified by his speech. In other words, numbers derive from God.

Imaging

How do numbers derive from God? Again, many perspectives are possible. But we can look at the question through the lens provided by the theme of imaging. The tabernacle is an *image* of God's dwelling in heaven. Within the tabernacle, the Most Holy Place is the most direct and intense image of God's presence. In it are (1) the ark, the most holy object of furniture, (2) the cherubim, who present an image of the cherubim in God's presence in heaven, and (3) the Ten Commandments, God's speech from heaven, which are deposited inside the ark (Ex. 25:16). The Holy Place is a less intense image. The priests are allowed to enter it every day, whereas the Most Holy Place can be entered only by the high priest, once a year. In some ways the Holy Place is like an image of an image: it "images" the Most Holy Place, by having the same dimensions in width and height (10 cubits), and being an exact proportion in length: 20 cubits, compared to the 10 cubits length for the Most Holy Place. Just as the Most Holy Place is a kind of dynamically constructed reflection and extension of heaven, the Holy Place is a kind of dynamic extension to the Most Holy Place. It has a derivative holiness, derived from being next to the greater holiness of the Most Holy Place.

20 cubits is 10 plus 10. The tabernacle as a whole, composed of the two rooms together, is 30 cubits long, or 20 + 10 cubits. We see simple arithmetical relationships. These relationships include the relationship of addition. The Holy Place is a kind of "addition" to the Most Holy Place, and the dimensions add to one another in a simple way.

This one example is a key example, because the tabernacle is an image for the whole universe as a large-scale house. By God's design, arithmetical relationships hold for the tabernacle. They can also be expected to hold for the universe as a whole. The relationships of 20 cubits = 10 + 10 cubits, and 30 cubits in length from 20 cubits + 10 cubits, are particular key examples. By generalizing from these examples, we confirm that by God's design arithmetical truths hold for the entire universe.

Where did all these designed harmonies come from? They came from God. They are "images," in the broad sense of the term, of God's dwelling in heaven. We know from New Testament teaching that the final dwelling of God is not simply his dwelling in heaven but his dwelling in Christ. "For in him [Christ] the *whole fullness of deity* dwells bodily" (Col. 2:9).

Origins in the Trinity

The word *bodily* shows that the verse in Colossians is focusing on Christ as the incarnate redeemer. But his role as incarnate redeemer presupposes his deity and therefore his eternality as the Word who was in the beginning (John 1:1). He always was with God. This eternal presence with God takes the form of indwelling. Jesus speaks of the fact that the Father is "in" him and he is "in" the Father (17:21). That mutual indwelling is the archetype for the dwelling that the Father and the Son will have in believers: "If anyone loves me, he will keep my word, and my Father will love him, and we will come to him and *make our home with him*" (14:23). This dwelling of God in man takes place through the Holy Spirit: "You know him [the Holy Spirit], for he *dwells* with you and will be *in* you" (v. 17).

These descriptions of indwelling come in the context of redemption. But when God acts to redeem us, he acts in harmony with who he really is, and he reveals himself to us in accord with who he is. Thus, the redemptive descriptions indicate not only that God exists in three persons, but that the three persons indwell one another. Theologians have given a name to this indwelling: *coinherence*.

Thus, the coinherence of persons in God is the archetype for God's dwelling in heaven, and then for the tabernacle and the temple. The tabernacle is an image of the archetype.

The origin of imaging is also found in God. The Son, the second person of the Trinity, is called "the image of God" (2 Cor. 4:4), "the image of the invisible God" (Col. 1:15), and "the radiance of the glory of God and the exact imprint of his nature" (Heb. 1:3). The Son as the original image is the archetype for the pattern when God says, "Let us make man in our image, after our likeness" (Gen. 1:26). Man is a subordinate or derived image, an image of an image. This pattern is similar to the tabernacle, which is the image of God's dwelling in heaven, which in turn is an image of God's dwelling in himself in the coinherence of the persons of the Trinity.

In fact, the language of sonship is closely related to imaging. When Adam fathers his son Seth, it is said, "he fathered a son in his own likeness, after his image, and named him Seth" (Gen. 5:3). Seth undoubtedly looked a little like his father, as most sons do, and he was like him in other ways as well. This likeness is one aspect of being a son to a father. Why? This pattern of sonship on earth is imitating (imaging!) the pattern of eternal Sonship in God. The Sonship that the Son enjoys in relation to the Father includes the Son being the image of the Father. When God created man on earth, he intended that the human relation between father and son would be an image of the eternal relation between God the Father and God the Son.

The father-son relation on earth is a dynamic one. Adam *fathered* a son, Seth. In the old-fashioned language of the King James Version, he "begat" a son. "Begetting" is fathering. This language applies to God the Father in relation to the Son. In Acts, the language of "begetting" applies to the fact that the Father raised the Son from the dead:

> What God promised to the fathers [patriarchs of Israel], this he has fulfilled to us their children by raising Jesus, as also it is written in the second Psalm,
>
> "You are my Son,
> today I have *begotten* you." (Acts 13:32–33)

God the Father's relation to the Son was also manifested earlier in time, when Jesus became incarnate:

> The Holy Spirit will come upon you [Mary], and the power of the Most High will overshadow you; therefore the child to be born will be called holy—the *Son* of God. (Luke 1:35)

As we have already seen, what God accomplishes redemptively expresses what he is in his character. So theologians have spoken of the eternal *begetting* of the Son. The conception of Jesus in Mary's womb took place in time. But it expresses in time an eternal relationship, which we cannot comprehend, but which we know is in accord with what happened in time when the Son became incarnate. The Father eternally begets the Son, expressing an eternal relationship between the two persons. The Holy Spirit is present, in coinherence, just as the Holy Spirit was present to "come upon" Mary.

The eternal begetting of the Son is also the eternal imaging, in which the Father begets the Son as his exact image. This imaging is the archetype, while other instances of imaging are ectypes.

This reality about God is relevant for tracing the origins of addition. A key instance of addition is found in the tabernacle and its rooms. One room, the Holy Place, is an addition to the original room, the Most Holy Place, through imaging. The original for this pattern is found in the imaging in the Trinity, which is also begetting.

We must here take care to underline the uniqueness of God. God is the Creator, and we are creatures. There is nothing like God. He is unique. The begetting and the imaging in God are therefore unique. But precisely in his uniqueness, his glorious uniqueness, God is the archetype for created, derivative patterns. Precisely because he is God, he can create a world distinct from himself, which reflects or images who he is. Addition, on the level of our earthly conception, exists because, first of all and primarily, the Father begets the Son in the presence of the Holy Spirit and in the love of the Holy Spirit. The Son is distinct from the Father, not the same. They are two persons.

It is precisely in accordance to his character, then, that God creates a world in which there can be an addition of an outer room of the tabernacle to an inner room. And we, as creatures, can think about adding one room to another, or one measurement to another. 10 cubits plus 10 cubits makes 20 cubits.

8

The Idea of What Is Next

As we have seen, the pattern for imaging begins in the Trinity, in that the Son is the exact image of the Father. The pattern could have started and ended there, because God did not have to create a world. But he did. In this world, he made more images of himself. There are multiple images. The most striking image is Adam, made "in the image of God." Adam fathered a son, Seth, in his image:

> When Adam had lived 130 years, he fathered a son *in his own likeness, after his image*, and named him Seth. (Gen. 5:3)

Seth fathered Enosh (v. 6), whom we may infer was made in the image of God and also in the image of Seth. Enosh fathered Kenan (v. 9). And the line continued.

We can also see a second kind of imaging process with the tabernacle. The Most Holy Place is an image of heaven. The Holy Place is an image of heaven and of the Most Holy Place. The tabernacle courtyard, surrounding the tabernacle, is also a holy space, and so is a kind of image of the Holy Place. Israel and Palestine and the holy land are images of the tabernacle and of its courtyard. And Israel was supposed to be a model to the nations, if they served God as they should (Deut. 4:6–8). It is clear that images can engender further images. The whole human race has come into existence by a process of repeated fathering, beginning with Adam.

Varieties of Succession

All these cases use the idea of "what comes next." Seth, the son of Adam, is next after Adam, and Seth's son is next after him. The process of imaging, by engendering a next thing, becomes a source for repeatedly increasing the number of things—the number of human beings, in one case, and the number of holy objects, in another.

This idea is clearly one of the ideas present when we view the numbers as a sequence. Each number has a next one after it: 2 after 1, 3 after 2, 4 after 3, and so on.

God exercised creativity in making a world when it was not a necessity for him. Seth had a son even though he might not have. These cases are analogous to the case with numbers, and we can view the numbers (from one of many possible perspectives) as summarizing a pattern of imaging or engendering. Numbers in their sequence represent the pattern of "what comes next" as a generalized pattern.

Because of the creativity involved, sequences of engendering can be of several types. The engendering could stop after the first replication: *A* to *B*. Or it could stop after four replications: *A* to *B* to *C* to *D* to *E*. Or one stage could father several later stages: *A* could produce B_1, B_2, B_3, and B_4. One of these second-stage *B* elements, let us say B_1, would then produce C_1, C_2, and C_3. C_2 produces D_1, and so on. The process as a whole produces a pattern like a genealogical tree going from a father to all his descendants. (See diagram 8.1.)

Diagram 8.1: Genealogical Tree

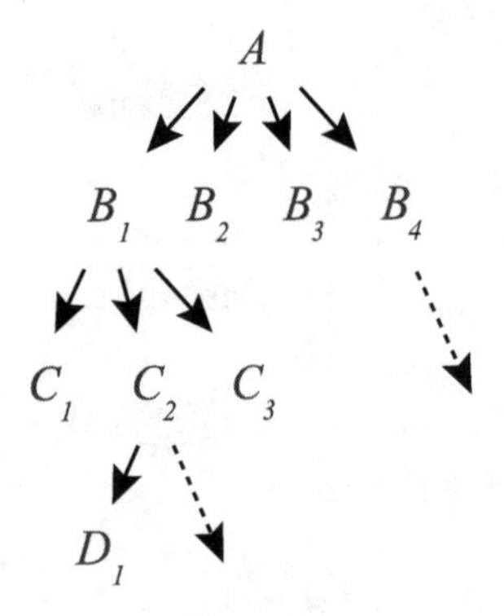

Among these possible creative choices, we may choose the simple one in which each item, once produced, imitates the previous production by producing one more item. Then we get a line of items:

A → B → C → D → E .

If we imagine the line proceeding indefinitely, we have represented the natural numbers as a line:

A → *B* → *C* → *D* → *E* → *F* → *G* → *H* → *I* → *J* → . . . .

1 → *2* → *3* → *4* → *5* → *6* → *7* → *8* → *9* → *10* → . . . .

If we imagine something unusual, like returning at some point to the first item *A*, we get a model that can be the starting point for what has been called "clock arithmetic" or "modular arithmetic." (See diagram 8.2.)

Diagram 8.2: Clock Arithmetic

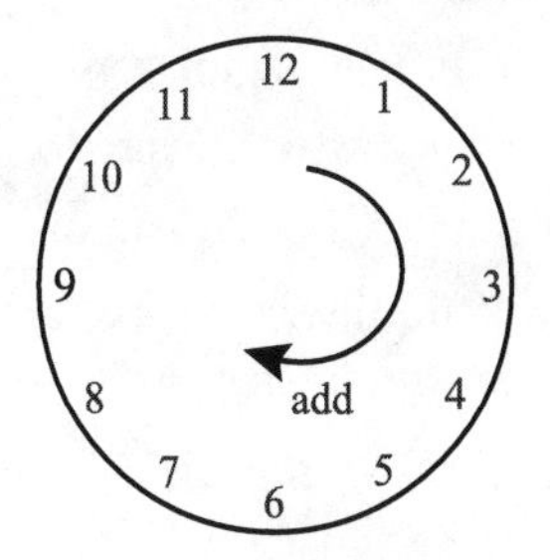

Some people might not want to call the system of hours on a clock "arithmetic" at all, since it differs from the familiar arithmetic. But the expression "clock arithmetic" explains intuitively what is happening. On a 12-hour clock, after we reach 12 o'clock, we can add one more hour and get back to 1 o'clock. 12 "+" 1 = 1 when we are moving around a clock. In the expression 12 "+" 1 = 1, I have put the plus sign + in quotation marks, to remind us that this "plus" symbol no longer has exactly the same meaning as it does in ordinary arithmetic. But there are some fascinating similarities to ordinary arithmetic, if we should choose to study how to add and subtract on clocks.

Natural Numbers

The simplest way of proceeding to what is next is to have one next thing that is new. We can always stop, and then we get a list of numbers, ending with the last, let us say 6. This list corresponds to a list of 6 apples or 6 distinct things. If we are thinking in terms of numbers rather than particular objects like apples, we are generalizing. That kind of generalization, as we have already seen, is one aspect of the meaning of 6. But now we are tying its meaning into the process of engendering or imaging. That is another aspect of its meaning. (God ordains all the aspects; we do not need to reduce them all to one.)

Rather than stopping at 6, we can imagine ourselves going on indefinitely. Do we ever get the entire list of whole numbers? No. We are finite, so we get weary or we run out of time. But we should notice that we have the capability, as people made in the image of God, of exercising a kind of miniature version of transcendence.[1] We can stand back from what we have been doing and look down on our previous actions, regarding them as a whole. In standing back, we imagine something of what it would be like to take a God's-eye view of what we have been engrossed in. We remain finite, but still we imitate God. We can imagine in some ways what it means for him to transcend the world, because we can in a miniature way "transcend" our surroundings and our immediate task. We have this gift from God. We are imitating him, though on the level in which we remain creatures.

So as we stand back from the process of creating a succession of numbers, we can see by our miniature transcendence that we could go on indefinitely. We could go on *forever*. We cannot literally go on forever in this life, but we can imagine an indefinite repetition of the process of engendering. We can do so because we are made in the image of God. By this imagination, we can understand what it means to be a natural number or a whole number. Mathematicians use the expression *natural number* to describe any number in the sequence that we imagine producing, starting with 1 and going on indefinitely.

[1] See further Poythress, *Logic*, chapter 45.

God's Thoughts

Our idea of the sequence of numbers is not independent of God. We are trying to think God's thoughts after him, analogically. God is the original thinker. Our thoughts never surprise him. He has already thought them. He knows the end from the beginning (Isa. 46:10).

We can infer, then, that God knows the natural numbers. He knew them all along, before he even created mankind. He reveals his thoughts to us as we study numbers.[2]

We can now return to the question of numbers within other universes that God might create. Might the system of natural numbers and the system of addition with respect to natural numbers be different in another world than what it is here? We have seen in the case of clock arithmetic that there is a sense in which there is more than one "system" of "arithmetic" even within this world. And the genealogical tree indicates that there are many possible "systems" of succession. But when people ask about numbers in another universe, they are probably not asking that kind of question. They are asking about natural numbers, the numbers that are familiar to us in ordinary arithmetic. We can also assume that they are not merely asking about alternate *notations* or symbol systems to *designate* numbers in written script. We can designate numbers using Roman numerals: I, II, III, IV, V, VI, VII, etc. Or we could spell out the names: one, two, three, four, five, ... These variations are interesting. But in another world, would the truths about numbers remain the same?

What we mean by natural numbers is closely connected to the concept of the *sequence* of natural numbers. And it is connected to our ability, by miniature transcendence, to see the numbers as an indefinitely extending sequence. This sequence is based on imitating the imaging or "engendering" process, which starts with 1 being succeeded by 2. We imitate the process again and again. All of this thinking is rooted in God, who is the Trinitarian God. God remains the same. So there is a stability and reliability to the number sequence. Arithmetic truths remain true always and everywhere. Because they have their foundation in God and in the Son who is the eternal image of God, we can conclude that the truths

[2] On God's involvement in knowledge of an ordinary sort, see ibid., chapter 15.

are necessary, once we follow along the path of thinking God's thoughts after him analogically.

But there is a caution. Our understanding of numbers is connected not only with the mind of God but with our experiences of four apples and four peaches. Apples and peaches might not exist in another universe, and the application of arithmetic truths to apples and peaches would not make sense if these particular fruits did not exist. The laws would still apply to grapes or pears if these fruits existed, or to foxes or blades of grass. We can still apply the truths if we *imagine* some new fruit that does not exist. But that is different. Because of the interlocking of the one and the many, we ought to resist the idea that we can completely separate the knowledge of numbers from the knowledge of ways in which they are illustrated in practice, within this world.

9

Deriving Arithmetic from Succession

The idea of succession can be used as a starting point to derive all of arithmetic. In 1889 Giuseppe Peano, building on the work of predecessors, treated the relationship of succession as fundamental. Starting with that relationship, he formulated precise axioms from which he could deduce all the elementary truths of ordinary arithmetic.

Peano's Axioms

The successor relationship can be described using a special symbol S. To express the fact that 2 is the successor of 1, we write that S1 = 2. Likewise, S2 = 3, S3 = 4, and S4 = 5. In order not to clutter the formulation of the axioms, we can assume that each natural number is named by using only the symbol *1* and the symbol *S* for successor. So the number 3 is SS1, with two occurrences of the symbol S. 4 is SSS1. And so on. For large numbers this notation becomes cumbersome. But the point is not to have an efficient notation, but to have simple axioms.

Here are the axioms:[1]

1. 1 is a natural number.
2. The successor of any natural number is also a natural number: that is, for all natural numbers *n*, *Sn* is a natural number.

[1] For technical completeness, the axioms would also have to include axioms describing the properties of the equality relation =. Peano's axioms are used in several forms, not all of which are logically equivalent.

3. No natural number has 1 as its successor: for all numbers n, $Sn \neq 1$.
4. No two natural numbers have the same successor: if $m \neq n$, $Sm \neq Sn$.
5. Suppose that M designates any property[2] that might or might not hold for a particular number n. Suppose that (a) M is true for 1 and (b) if M is true for a number n, it is also true for Sn. Then M is true for all natural numbers.

These five axioms, simple as they appear to be, can be used to define addition and multiplication, and then to deduce all the elementary results of ordinary arithmetic (see appendix C). It is an impressive achievement to have found a way of representing arithmetic in such a simple way. There is a beauty to the simplicity of the axioms, and naturally this beauty is a reflection of God.

We have already discussed the fact that numbers enjoy a multitude of relationships with many aspects of life (chapter 4). The relationship to Peano's axioms is one such relationship. The usefulness of the axioms does not mean that numbers are *reduced* to the axioms; rather, given our antireductionist philosophy, it means that numbers enjoy logical relationships to these axioms, or to other axioms that we might pick. There are many possible choices of axioms that would lead to the same results. Peano's axioms are in some ways the simplest. But they enjoy relationships to other axioms. For example, one possible alternative set of axioms starts with zero rather than one as the lowest number. Axiom 3 then has to be adjusted to say that no number has zero as its successor. All the other axioms are the same. This new system of axioms results, of course, in a slightly different definition of the natural numbers, since with the new set of axioms zero is included among the natural numbers. But the properties of the numbers are the same. We could also pick a set of axioms in which addition and multiplication are already defined. All the possibilities for different axioms reside in the mind of God before we

[2] There are complexities about what is allowed as a "property" M. If we allow only properties that can be expressed using a formal logical language with first-order quantification, we have enough to establish many elementary truths of arithmetic, but not enough to define uniquely everything about natural numbers. We must leave such issues to more advanced books about the axiomatization of arithmetic.

as human beings start thinking about them. They enjoy relationships to alternative sets of axioms, in accordance with the wisdom of God and his self-consistency. In this area, as in all others, we can praise God for his wisdom and richness.

In addition, we can observe that if we are going to understand Peano's axioms properly, we already have to know about numbers. Truths about numbers can be derived from Peano's axioms, but Peano's axioms can also be derived from truths about numbers. The successor relationship can be seen as a special case of addition: S*n* for any number *n* can be seen as an alternate notation for the concept of addition by 1 that we already have in mind. S*n* is shorthand for $n + 1$. Or consider Peano's notation for the number 4, namely SSS1. We have to be able to count the number of occurrences of the symbol *S* in the expression. So we are already dependent on numbers and on counting when we start.

The Foundations for Peano's Axioms

We can consider Peano's axioms one by one, and reflect on ways in which they have foundations in the character of God. Let us begin with axiom 1:

1. 1 is a natural number.

This axiom makes sense in a world in which God has ordained the patterns for arithmetic truths. These patterns have their archetype or origin in God's self-consistency. The number 1 has its archetypal origin in the unity of God, who is one God.

2. The successor of any natural number is also a natural number: that is, for all natural numbers *n*, S*n* is a natural number.

Axiom 2 has its roots in the idea of succession or "what is next." We have indicated in the previous chapter how this idea has its roots in imaging and "engendering," which go back to the Son, who is the original, archetypal image, begotten by the Father in an eternal begetting.

3. No natural number has 1 as its successor: for all numbers *n*, $Sn \neq 1$.

Axiom 3 specifies that we are not dealing with clock arithmetic. The succession of numbers never circles around to come back to the beginning. This principle follows when each round of producing a successor imitates the first round by producing a *new* successor rather than repeating an older one. The idea of newness goes back to the newness that took place when God created the world.

4. No two natural numbers have the same successor: if $m \neq n$, $Sm \neq Sn$.

Axiom 4, like axiom 3, specifies that each new successor is indeed genuinely new.

5. Suppose that M designates any property that might or might not hold for a particular number n. Suppose that (a) M is true for 1 and (b) if M is true for a number n, it is also true for Sn. Then M is true for all natural numbers.

Axiom 5 is clearly a key axiom, because it implicitly involves the entire succession of natural numbers. It is called the axiom of *mathematical induction*. The process of mathematical induction is a form of reasoning that starts with a particular property M, and wants to show that it is true for all natural numbers. It establishes the general truth about M by going through the two steps (a) and (b). The steps (a) and (b) are sufficient, because, using these steps, we can see how the property can be established for each natural number in succession.

Let us see how it works. Suppose that steps (a) and (b) are true for a particular property M. We reason as follows:

1. M is true for 1 (by step (a)).
2a. If M is true for 1, M is true for $2 = S1$ (by step (b)).
2b. M is true for 2 (from lines 1 and 2a).
3a. If M is true for 2, M is true for $3 = S2$ (by step (b)).
3b. M is true for 3 (from lines 2b and 3a).
4a. If M is true for 3, M is true for $4 = S3$ (by step (b)).
4b. M is true for 4 (from lines 3b and 4a).
5a. If M is true for 4, M is true for $5 = S4$ (by step (b)).
5b. M is true for 5 (from lines 4b and 5a).

By repeating this process a sufficient number of times, we can establish that *M* is true for any natural number that we choose. The distinct element in axiom 5 is to say that then it is true not just for a particular number that we choose, but for all natural numbers whatsoever. That conclusion is possible only if we see the overall pattern. We stand back from the process of reasoning, and see a general pattern. We extrapolate the pattern forward along the series of numbers, and we see that the principle encompasses all of them. In the process of reasoning, we have used our ability to have miniature transcendence, to see a whole even when it is indefinitely extended. We are imitating the mind of God. We are finite, but with this kind of projection forward we depend on his infinity.[3] (For examples using mathematical induction, see appendices C and D.)

Thus all of Peano's axioms reflect the wisdom and greatness of God, each axiom in its own way. When we reason about arithmetic, we reason in imitation of God's prior knowledge of all truth, including arithmetical truth. Praise the Lord!

We may also observe that these axioms are in harmony with all the individual truths of arithmetic, and that both axioms and individual truths are in harmony with the world that God has made, where two apples plus two apples equals four apples. And all these areas together are in harmony with human minds, because we are made in the image of God. Because we share his image, we can teach the next generation to know truths in harmony with what we know and in harmony with what God knows. God is in harmony with himself, and ordains a world that reflects his harmony.

[3] For mathematical induction, see also Poythress, *Logic*, chapter 45.

10

Multiplication

Like addition, multiplication is established by God and originates in God.

Proportions in the Tabernacle and the Temple

We can return to consider the tabernacle of Moses and the temple of Solomon. The rooms in the tabernacle and in the temple show simple proportionalities in their dimensions. The Holy Place in the tabernacle is 10 × 10 × 20 cubits, in comparison to the Most Holy Place, 10 × 10 × 10 cubits. The Holy Place can be viewed as an image or addition to the Most Holy Place, giving us an example of addition. It can also be viewed as a room obtained, figuratively speaking, by multiplying the Most Holy Place by two in length. Simple proportions show a harmony. This harmony reflects on the created level the eternal harmony among the persons of the Trinity.[1]

The tabernacle and the temple both show multiple patterns. The relationships between the Most Holy Place and the Holy Place show numerical patterns, as we have seen. They also show spatial patterns. The Most Holy Place is a perfect cube, with length, breadth, and height all 10 cubits. These dimensions make up a space in which there can be patterns of motion and human activity, as a priest enters and performs his duties. The furnishings show patterns of physical support. And the tabernacle displays beauty in its furnishings. The multiple aspects, such as

[1] See further Poythress, *Redeeming Science*, chapters 20–22.

the numerical, the spatial, the physical, and the beautiful, all combine into a single structure.

This combination of aspects occurs also in the larger world that God made. Quantitative and spatial aspects of the world belong together with many other aspects, according to God's design. The implication is that God in his wisdom has made the world a whole. The combination into one whole as well as the individual aspects when contemplated separately display his wisdom. Quantitative and spatial aspects, which form the stuff for mathematical reflection, belong together with many other aspects. Mathematics is not more ultimate or less ultimate than many other aspects. Thus we should not be tempted either to glorify mathematics or to despise it as unimportant.

We can also note that multiplication is closely related to addition. Adding a number to itself is equivalent to multiplying the number by two: $3 + 3 = 3 \times 2 = 6$. Adding a number to itself for a total of four occurrences of the number is equivalent to multiplying by 4:

$$3 + 3 + 3 + 3 = 3 \times 4 = 12.$$

This property can even be used as a definition of multiplication, if we like (see appendix C). Or we can use a perspective where we start with addition and multiplication as distinct operations, and then show that they interlock harmoniously.

Multiplication in the World That God Made

Multiplicative properties find embodiments and illustrations in many ways in the world that God has made. For example, the area of a rectangle is the length multiplied by the width (diagram 10.1).

Many basic physical laws involve the mathematics of multiplication. One of the most famous laws is Newton's second law of motion,

$$F = ma\,.$$

It says that the force F is equal to the mass m *multiplied by* the acceleration a. Einstein's famous equation

$$E = mc^2$$

says that energy E is equal to mass m *multiplied by* the speed of light c *multiplied by* c a second time (the square).

Diagram 10.1: Area

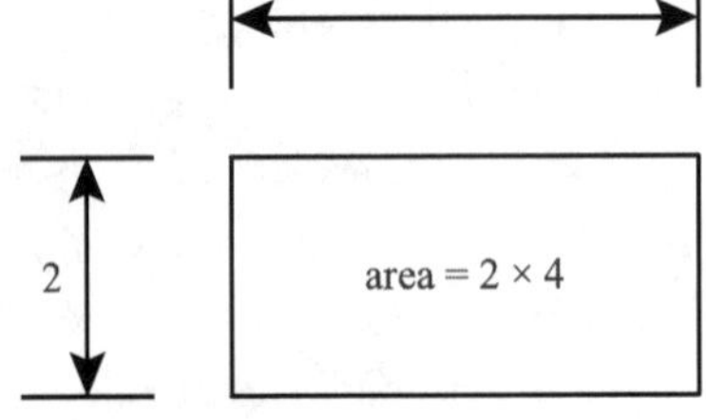

Multiplication of Animals

Genesis 1 describes God's ordering of the world. Among his commands, he specifies that animals and mankind should *multiply*:

> And God blessed them, saying, "Be fruitful and *multiply* and fill the waters in the seas, and let birds *multiply* on the earth." (v. 22)

> And God said to them [human beings], "Be fruitful and *multiply* and fill the earth and subdue it, and have dominion . . ." (v. 28)

Of course, the verses here are not describing "multiplication" in a technical mathematical sense. The meaning is more general: God is directing the animals and human beings to increase in number. But when we pay attention to *how* this increase takes place, we discover a relationship to multiplication in the mathematical sense. A species does not increase merely by each pair producing a single new offspring. A single pair can produce several offspring, and these offspring in turn may each produce several offspring.

Suppose we simplify, and picture a situation where a pair of horses produces four offspring. The second generation has four horses, twice as many as the first generation. If these four horses pair up, each of the pairs can produce four offspring in the third generation, for a total of eight horses in the third generation. There are twice as many in the third generation as in the second. If we repeat the pattern, we can have twice eight or 16 horses in the fourth generation. In the 10th generation there would be 1,024 horses, and in the 20th generation 1,048,576. In the 30th generation there could be 1,073,741,824, over a billion horses. The numbers become huge. We can see a dramatic difference between this kind of multiplication and a simple process of addition where, say, we add one new horse for each generation.

The pattern of reproduction that God has established reflects mathematical truths that have their foundation in God. Among these truths are truths concerning multiplication. We can see that truths about multiplication are integrally related to the normative, situational, and existential perspectives. First, the normative perspective leads to focusing on truths of multiplication as general truths that hold for horses, cats, or any other objects that we choose to count. Multiplication works in general, and we can logically derive truths of multiplication from simple starting axioms (see appendix C). Second, the situational perspective leads to focusing on how multiplication works with horses, cats, and other objects. Third, the existential perspective leads to focusing on our capability as persons to understand how multiplication works and how it is significant. The three perspectives harmonize, according to God's design.

We can thank God that in this and many other ways he gives ability to human beings not only to understand the wonders of his world, but to use regular arithmetical patterns for our benefit, as when we undertake to breed animals.

11

Symmetries

Within mathematics we can find many symmetries. Let us reflect on a few of them.

What Is Symmetry?

A *symmetry* is displayed whenever one kind of change in viewpoint leaves something fundamentally the same. For example, the human body has *bilateral* symmetry: a person looks about the same in a mirror, even though the mirror reverses the positions of the left and right sides. Left and right eyes correspond; left and right hands correspond; left and right legs correspond. (See fig. 11.1.)

Fig. 11.1: Symmetric Face

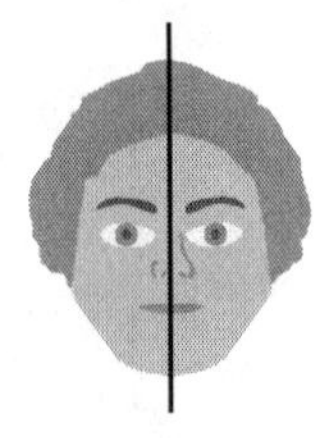

By contrast, a starfish has what is called a *radial* symmetry. A starfish has five arms, all of which are about the same shape. So rotating the starfish around its center by 1/5 of a complete revolution leaves the starfish looking about the way it did before. A starfish has in addition *mirror*

symmetry, shown by the fact that it looks about the same in a mirror. (See fig. 11.2.)

Fig. 11.2: Symmetric Starfish

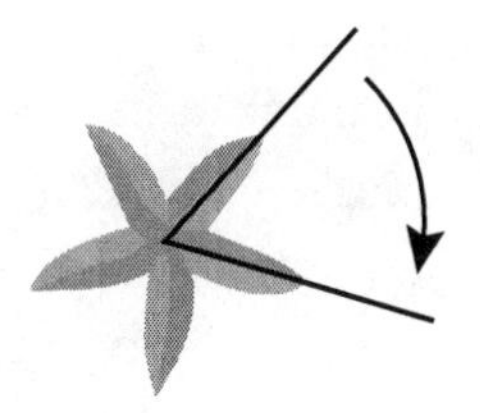

An earthworm has a cylindrical symmetry, so that it looks about the same after any amount of rolling. (See fig. 11.3.)

Fig. 11.3: Symmetric Cylinder

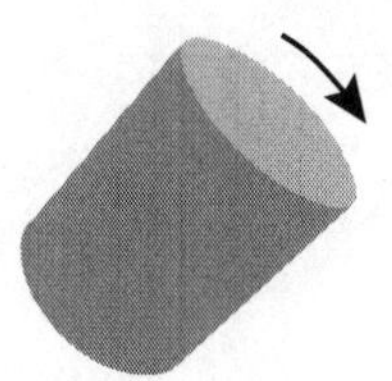

The cells in a honeycomb show a sixfold symmetry. A rotation by an angle of 60 degrees or any multiple of 60 degrees leaves the structure the same. The cells also look the same in a mirror. (See fig. 11.4.)

Fig. 11.4: Symmetric Honeycomb

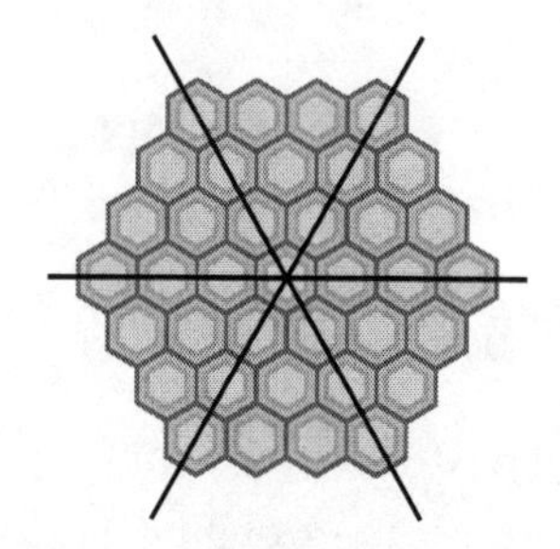

The Origin of Symmetry in God

Symmetries within this world exist because of God's plan. He made them. Within God himself, there is an archetype for symmetry, namely the fact that all three persons of the Trinity are equally God. The three persons are distinct from one another, and their roles are distinct in relation to one another, but they all share the characteristics of God—eternality, omnipotence, omniscience, omnipresence, faithfulness, goodness. They are "symmetric" with respect to these characteristics. This symmetry is the archetype. All earthly symmetries are ectypes, created reflections.

Symmetry has a close relation to beauty. People tend to think that a human face with symmetry is more beautiful than one that lacks symmetry at some point. The beauties in this world reflect the archetypal beauty of God.

Symmetries in Arithmetic

Arithmetic shows simple symmetries. Each of these symmetries ultimately reflects the beauty of God.

Addition is *commutative*, meaning that the order of two numbers makes no difference:

$$1 + 3 = 3 + 1 = 4;$$
$$2 + 3 = 3 + 2 = 5;$$
$$5 + 7 = 7 + 5 = 12;$$
$$6 + 9 = 9 + 6 = 15.$$

(For a demonstration of commutativity, see appendix C.) This property is a symmetry with respect to the order in addition.

Addition is *associative*, meaning that the grouping of three numbers makes no difference:

$$(1 + 2) + 4 = 1 + (2 + 4);$$
$$(2 + 1) + 5 = 2 + (1 + 5);$$
$$(7 + 3) + 2 = 7 + (3 + 2).$$

(For a demonstration of associativity, see appendix C.) This property is a further symmetry with respect to grouping in addition.

Together, the commutativity and associativity in addition imply that the order of addition makes no difference even with a long sequence of numbers to add. (See diagram 11.1.)

Diagram 11.1: Addition Harmony

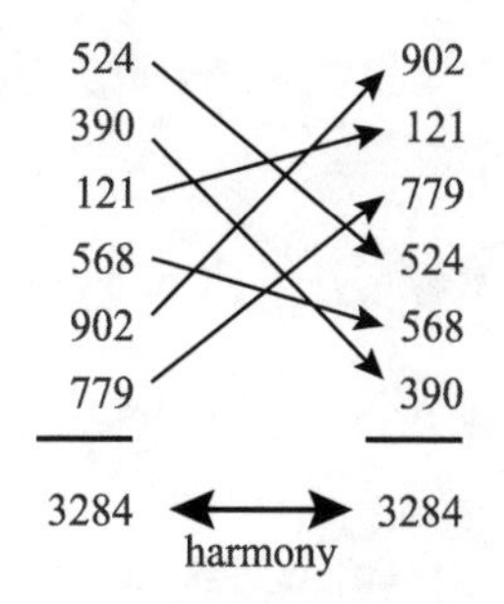

Thus there is a complex symmetry in the fact that a rearrangement of order leaves the sum the same.

Multiplication is commutative:

$2 \times 3 = 3 \times 2 = 6;$
$3 \times 4 = 4 \times 3 = 12;$
$3 \times 7 = 7 \times 3 = 21;$
$5 \times 11 = 11 \times 5 = 55.$

This commutativity is a symmetry for multiplication.

Multiplication is associative:

$(2 \times 3) \times 4 = 2 \times (3 \times 4) = 24;$
$(2 \times 5) \times 3 = 2 \times (5 \times 3) = 30;$
$(3 \times 6) \times 4 = 3 \times (6 \times 4) = 72.$

We can appreciate all these symmetries as beauties that God has placed within arithmetic for our enjoyment. Many more complex symmetries and beauties await us in mathematics. The further one travels in the study of mathematics, the more there are, and the more we should be stimulated to praise God. An excellent resource is found in James Nickel, *Mathematics: Is God Silent?*[1]

[1] James Nickel, *Mathematics: Is God Silent?* (Vallecito, CA: Ross, 2001).

12

Sets

Sets are widely used in mathematics. And some people have tried to reduce all of mathematics to the theory of sets. So we need to reflect on the nature of sets as mathematical objects.

What Is a Set?

A *set* is a collection of objects. We can indicate what set we have in mind simply by listing the objects that are in the collection. For example, the collection consisting of apple #1, apple #2, and peach #1 is a set. When there are only a few items in the collection, we can describe the set by putting the list of items inside braces. The set S that consists in three objects, apple #1, apple #2, and peach #1 is described thus:

$S = \{\text{apple \#1, apple \#2, peach \#1}\}.$

The objects in a set are called *members* or *elements* of the set. The symbol ∈ is used to denote set membership.[1] The expression

$(\text{apple \#2}) \in S$

means that apple #2 is a member of the set S.

A set is a collection in which we ignore all the information except the information about what objects it has as its members. The technical con-

[1] The symbol ∈ is a form of the Greek letter epsilon. But as a unicode character it is distinct from the Greek alphabet. Unicode characters U2208 and U220A are both used for this purpose.

cept of *set* is "blind" to all the extra information and extra relationships and associations that we have in our minds from our life as a whole. So the set is the *same* set, no matter what order we choose in which to list the elements, and no matter how many times we list the same element:

{apple #1, apple #2, peach #1} = {peach #1, apple #1, apple #2}
= {apple #2, apple #2, apple #2, apple #1, peach #1, apple #2, apple #1}.

The idea of a set singles out by mental abstraction just those properties on which we are focusing, and for sets the key property is membership in the set. The general properties of a set that make it a set represent the abstraction, while particular instances of sets, like the set {apple #1, apple #2, peach #1} represent concrete embodiments or illustrations of the abstraction. The abstraction is one in nature, while the embodiments are many. Because of the equal ultimacy of the one and the many, the abstraction and its embodiments belong together, each helping to define the other.

If there are many members to a set, it may be more efficient to describe the set by describing the properties that are true of every member of the set. For example, the set of all odd numbers less than 100 can be described in ordinary English just as we have done. There is also a standard way that mathematicians have for writing out the description in an abbreviated form:

the set of odd numbers less than 100 = $\{x \mid x$ is a natural number and $x < 100$ and x is odd$\}$

The bar symbol "|" means "such that." The notation $\{x \mid x$ is odd$\}$ means "the set of all elements x such that x is odd."

Foundations for Sets in God

As usual, the foundation for the idea of sets is in God. Let us see how. The idea of a set depends on three principles: (1) each element in the set is distinguishable and fixed—it is well defined; (2) there is a clear criterion for distinguishing which objects are in the set and which are not; (3) there is a relationship of belonging or "membership," according to which the elements that meet the criterion for being in the set are said to be *members* or *elements* of the set. The relationship of belonging is denoted by the special symbol ∈.

The principles (1) and (2) both depend on the possibility of making distinctions. How can we make distinctions? God is distinct. He is one God. And God is three persons. The three persons are distinguishable from one another. At the same time, the three persons belong together: each person is God. Each person, we might say, is a *member* of the Godhead. We saw in chapter 2 that the three persons of the Trinity give us the archetype for three principles: classification, instantiation, and association. The principle of classification is the archetype for criteria for distinctions. The principle of instantiation is the archetype for the individuality that belongs to elements that meet or do not meet criteria. And the principle of association is the archetype for the relationship of belonging or membership. All three principles interlock. We cannot really have one without the others. All three presuppose the others, in harmony with the fact that all three persons of the Trinity belong together as one God.

God, we said, is the archetype. When God created the world, he created ectypes that reflect his wisdom, his faithfulness, and his knowledge. In the description in Genesis 1 we can see, among other things, that God makes *distinctions*:

> And God *separated* the light from the darkness. God *called* the light Day, and the darkness he *called* Night. (Gen. 1:4–5)
>
> And God said, "Let there be an expanse in the midst of the waters, and let it *separate* the waters from the waters." And God made the expanse and *separated* the waters that were under the expanse from the waters that were above the expanse. And it was so. And God *called* the expanse Heaven. (vv. 6–8)
>
> And God said, "Let the waters under the heavens be *gathered together* into *one* place, and let the dry land appear." And it was so. God *called* the dry land Earth, and the waters that were gathered together he *called* Seas. (vv. 9–10)

> And God said, "Let there be lights in the expanse of the heavens to *separate* the day from the night. And let them be for signs and for *seasons*, and for *days* and *years*, . . ." (v. 14)

Distinctions among the things that God has made come about because God speaks the distinctions into existence. The names that he gives, such as Day, Night, Heaven, and Earth, define distinct elements within the world. He *separates* things, which has the result that the things separated are then distinguishable. Among other things, these distinguishable things can then function as distinct elements within a set.

Genesis 1 is describing God's work of creation by way of overview. But this overview, by illustrating the use of distinctions, implies a more general principle. God specifies all the distinctions that exist. We as human beings can think God's thoughts after him. When we do so, we rely on distinctions that are already in place, because God specified them.

In our day, there are many languages of the world. They include differing vocabularies, and the vocabularies may sometimes focus on different *kinds* of distinction. The vocabularies do not always match one another. But whatever distinctions exist in whatever languages, God has thought about them beforehand.

The idea of a set utilizes distinctions in a simple, clean way. It strips away all other kinds of detail about the things in God's world and presents the simple idea of being a member of a set or not. An item is either inside the set, by being a member, or outside the set, by not being a member. That inside-outside distinction is a separation. We draw up or "create" the separation when we define a particular set with apple #1, apple #2, and peach #1 as its members. We are "creative." But of course our creativity is derivative. We are made in the image of God our Creator. God has thought through all the distinctions and separations before we did.

Perspectives on Sets

Sets can be understood using the normative, situational, and existential perspectives. The normative perspective focuses on the normative properties of sets. Particular truths hold for sets. These truths have a transcendent source—ultimately they come from God's self-consistency, and they depend on the fact that God has specified distinctions and separations.

Second, the situational perspective focuses on the situation: objects in the world that God has created. We can treat these objects, like apples and peaches, as members of collections. Truths about sets hold for such collections in the world.

Third, the existential perspective focuses on persons. We as human beings have to be able to grasp the meaning of talking about a collection, or talking about its members. We have to be able to use our minds to think about distinctions, and to be clear in our minds as to what distinctions we are using at any time.

As usual, these three perspectives lead to one another. In our minds we think about the world. So the existential perspective leads to the situational perspective. And when we think about the world, we presuppose norms, including norms for what is true of sets. That is, our thinking includes awareness of norms, and so we are led to the normative perspective. The norms hold true for things in the world, and so the normative perspective leads to the situational perspective. The norms hold true for how we should think, and so the normative perspective leads to the existential perspective.

The three perspectives harmonize because God has ordained them all. He specifies the norms; he creates the world; and he creates human beings in his image.

Sets and Numbers

Are sets part of the foundation for numbers? As we indicated, it is common for contemporary mathematicians to consider set theory as a "foundation" for everything else. What this often means is that mathematicians can start with axioms for sets, and "build up" or derive all the truths about mathematics from this starting point. Our viewpoint here is antireductionistic. The derivability from axioms displays the fascinating relationship between sets and numbers, and displays God's wisdom in ordaining a relationship. But the relationships are rich and multidimensional.

For example, we are already tacitly using what we know about numerical features of the world when we think about sets. Each element of a set is *one* element. The distinctiveness of the element already presupposes the number one. If there is more than one element in the set, the distinction between elements gives us two or three elements in the set. We know this even if we do not mention it.

We cannot think about sets without thinking God's thoughts after him. We must have a kind of communion with God, even if we are morally and spiritually in rebellion against him. In this communion with God, we already know something about numbers.

We can also see that in God himself we find numbers (one God; three persons) and distinctions (distinctions between persons). Neither is "prior to" the other, since both belong to God from eternity. Numbering presupposes distinctions between the persons that we number. Conversely, distinctions presuppose the idea of unity and diversity, which is numerical.

So we can look at the subject in at least two opposite ways. On the one hand, the idea of distinction depends on the "prior" idea of numbers. Or we can see numbers as depending on the "prior" idea of distinction. Both actually go together. In appendix E we give a brief picture of one direction of dependence. Numbers can be seen as depending on the idea of distinction that is present in the idea of a set. So we can explore how elementary set theory can provide axioms that lead to the properties of numbers.

Part IV

Other Kinds of Numbers

13

Division and Fractions

The whole numbers 1, 2, 3, 4, … are easiest to understand, because they apply to collections of apples or peaches. But there are other kinds of numbers, such as fractions. What is the nature of fractions? The mathematician Leopold Kronecker is alleged to have said, "God made the integers; all else is the work of man." Is that so? Or did God give us fractions and other kinds of numbers as well?

Division

Fractions are useful when we have to deal with dividing up some quantity. So let us think about division. Division undoes the result of a multiplication. So consider a case involving multiplication. Suppose the grocery store has packages containing 12 hotdogs. We buy 3 packages. How many hotdogs do we have? The principle of addition says that we can add up the hotdogs in each package, for a total of $12 + 12 + 12 = 36$ hotdogs. Adding 12 to itself for a total of three occurrences of the number 12 is the same as multiplying 12 by 3.

$$12 + 12 + 12 = 12 \times 3 = 36.$$

What we are seeing so far is harmony ordained by God between addition and multiplication, and harmony between the arithmetic on the one hand and the nature of the world (the world with its hotdogs) on the other hand.

Now let us pose a problem that requires thinking in the other direction. Suppose we are planning a party with 18 guests, and we estimate that they will eat two hotdogs each. How many packages do we have to buy at the store? We first do a multiplication: 18 guests times 2 hotdogs per guest is 36 hotdogs. If the hotdogs come in packages of 12, how many packages do we need to buy? With 12 hotdogs per package, one package will give us 12 hotdogs; 2 packages will give us $12 \times 2 = 24$; 3 packages will give us $12 \times 3 = 36$ hotdogs; and so on. Getting an answer to the problem involves "undoing" the multiplication problem $12 \times 3 = 36$, to conclude that 3 packages are enough to provide 36 hotdogs for the party.

This kind of problem crops up frequently, so people have invented a notation for division: $36 \div 12 = 3$, or $36/12 = 3$. When we analyze division, we can see that it displays normative, situational, and existential aspects. In the normative perspective, there are rules for carrying out division, and rules for the relationship between division and multiplication. There are also rules that involve harmonies between division and addition. For example, dividing first by one number, then by another, is the same as dividing by the product of the two numbers:

$$(36/3)/4 = 36/(3 \times 4) = 3$$
$$(20/2)/5 = 20/(2 \times 5) = 2.$$

Dividing a sum of two numbers by a number d has the same result as dividing each of the original two numbers in the sum by the same number d, and then adding:

$$36/3 = (12 + 24)/3 = (12/3) + (24/3) = 4 + 8 = 12.$$
$$10/2 = (8 + 2)/2 = (8/2) + (2/2) = 4 + 1 = 5.$$

In the situational perspective, division applies to situations in the world, like the situation with our 36 hotdogs.

Existentially, we as human beings can understand in our minds the hotdog problem, and carry out a process of division that leads us to an answer. Once we have the answer, we then proceed to interact with the world by purchasing the hotdogs and—along with the guests—eating them. As usual, God ordains the harmony between the three perspectives.

We can also observe the presence of interlocking between one and many. The one in this case is the general truth that $36 \div 12 = 3$. The many are the many instances in the world where this numerical relationship is exhibited—with hotdogs, hotdog buns, hamburger patties, chicken legs, and so on. As usual, the interlocking of one and many depends on God and has its archetype in God.

We may also observe that there is a kind of symmetry between division and multiplication. We have said that division "undoes" multiplication. The two operations of multiplication and division are two sides of one coin. If division undoes multiplication, multiplication also undoes division. If we have divided 36 by 12 to get 3, we can get back to 36 by multiplying 3 by 12. Symmetry in this world derives from God, who is the archetype for beauty.

Another symmetry about division arises because division can be viewed from two different perspectives.[1] Consider again the hotdog problem where we know that we need a total of 36 hotdogs. When we purchase them, they come in packages of 12. Now $36/12 = 3$. So we know we need to buy 3 packages. But suppose that our problem is that we have 36 hotdogs and 12 people who will eat them. How many hotdogs will each person eat? We obtain the answer in the same way: $36/12 = 3$. Each person will have 3 hotdogs.

The two problems have the same solution in arithmetic. In both cases, we have to divide 36 by 12. But in the world of hotdogs and packages and people, the two problems are quite different. The purchase problem involves dividing 36 hotdogs into piles of 12 hotdogs each, and asking how many piles there will be. The eating problem involves dividing 36 hotdogs into 12 piles, one pile per person, and asking how many hotdogs each person will get. The two problems have the same numerical answer, namely 3. This sameness is a kind of symmetry in the world, a symmetry about dividing 36 hotdogs into 12 piles or dividing them into piles each of which consists in 12 hotdogs.

The same, of course, is true in other cases of division. 20 hotdogs divided into 5 piles results in each pile having 4 hotdogs ($20/5 = 4$). 20 hotdogs divided into piles with 5 in each pile results in 4 piles ($20/5 = 4$).

[1] I owe this insight to Gene Chase.

We can see that this symmetry will always be there if we pictorially represent the division problem by means of a rectangular arrangement of hotdogs (fig. 13.1).

Fig. 13.1: The Hotdog Problem

The hotdogs are arranged in four rows of five hotdogs each, for a total of 20 hotdogs ($5 \times 4 = 20$). If each row is a "pile," we have four piles with five hotdogs in each pile. If, on the other hand, each column is a pile, we have five piles with four hotdogs in each pile.

Fractions and Division

Now let us consider fractions. Suppose that we have one pie that we want to divide among six people. This is a problem similar to the problem of dividing up 36 hotdogs into packages of 12. But we start with only one item, the one pie, rather than 36. The answer is that we cut up the pie. God has given us power to cut up things that are in the world. And he has given us minds that can think through how to do it so that the resulting pieces are about the same.

We divide the pie into six pieces. Once the pie is in pieces, we can adopt a new perspective in which we treat the pieces as individual objects, and the whole pie as a collection of 6 pieces. (Our ability to use multiple perspectives comes from God.) If we have 6 pieces, and we want to divide them up equally among 6 people, how do we do it? We need to divide the total of 6 pieces by the number of people, namely 6 people. 6 pieces divided by 6 people is 6/6 or 1. Each person will get 1 piece.

But now we can also return to the original perspective, where we regard the pie as a single whole. The pie is 1 item. How much does each person out of the 6 get? He gets 1/6. Orally, we say "one sixth." Writing it or saying it that way extends the notation of division into fractions.

From one point of view, it is we who have created this extended notation. We have "invented" the fraction 1/6 in order to enhance our ability to talk about the process of cutting things up. There is a grain of truth in the statement attributed to Kronecker, that "all else is the work of man." Mankind is creative, and we "invented" fractions. But who gave us our creativity? We are made in the image of God, who is the original Creator. God is not surprised when we come up with the idea of fractions. He thought of it before we did. And he made certain things within the world that divide up naturally into smaller pieces. For example, after an orange is pealed, it divides up naturally into sections. Clam shells divide naturally in two.

Fractions display the same interlocking of three perspectives that we observed with hotdogs. Fractions are *not* a merely subjective invention to entertain us or keep our minds busy with some frivolity. To be sure, we understand fractions *mentally*: that is the focus of the existential perspective. But we also know that there are norms for dealing with fractions correctly. We will have too few pieces, or else too many pieces, if we calculate mistakenly when we undertake to cut the pie. (Think of cutting up three wedding cakes into pieces for 200 guests. We had better do our arithmetic correctly, or we may be embarrassed by not having enough pieces for all the guests.) There are norms for success, and these norms are the focus of the normative perspective. Finally, in the situational perspective we focus on the pie. It has to be cut up.

The three perspectives cohere because God has ordained all three, and he has made sure that they cohere. That is why we can cut up a pie in a reasonable way.

If God has ordained the three perspectives on fractions, it is a mistake to reduce the three perspectives to one, namely the existential perspective. If we thought that the existential perspective was ultimate, then we might conclude that fractions are wholly "the work of man." That is, we have just invented them in our minds, and they exist only because we invented them. But that exclusive claim about human invention does not explain why fractions work well when we are cutting up pies or wedding cakes. Nor does it explain why we cannot just invent any rules that we wish for working with fractions. The rules are norms. They have to be what they are, if they are going to match what is true for the world (the situational perspective) and what is true for our minds (the existential perspective).

The norms for fractions in many ways match the norms for other forms of division. For example, a fraction of a fraction has as its denominator the product of the denominators in the two steps of making a fraction:

$$(3/8)/4 = 3/(8 \times 4) = 3/32.$$
$$(1/3)/5 = 1/(3 \times 5) = 1/15.$$

The addition of fractions satisfies a "distributive" law that is similar to what takes place with multiplication and division of whole numbers:

$$(3 + 4)/8 = (3/8) + (4/8).$$
$$(1 + 3)/6 = (1/6) + (3/6).$$

These truths are similar to:

$$(3 + 4) \times 8 = 3 \times 8 + 4 \times 8$$
$$(1 + 3) \times 6 = 1 \times 6 + 3 \times 6$$

In all this reasoning, whether from a normative, situational, or existential perspective, our minds are not working in independence of God. God is present with us. He is present for salvation with those who believe in Christ. But he is also present in common grace with those who rebel against him. They too can think God's thoughts after him. The "invention" of fractions is an invention empowered by God. It is not "merely" human.

Rules for Fractions

When children learn to deal with fractions, they should be learning multiple relationships and multiple aspects. Fractions have relationships with the world in which we cut up pies. They have relationships to our minds. They have relationships to language, and especially to the mathematical symbols that we use in "doing" fractions on paper. They have relationships to calculations done in the sciences. They have relationships to various kinds of advanced mathematics that have beauties of their own, but not everyone needs to learn them. Appreciating fractions means appreciating a rich world of relationships that God has ordained.

Within that context, children learn norms—rules. There are informal rules for the way in which written fractions relate to the world of cutting up pies. There are more formal rules for calculating with fractions. How do we multiply two fractions? How do we add two fractions with the same denominator (1/7 + 3/7)? How do we add two or more fractions with different denominators (1/6 + 1/2 + 1/9)? These rules must have coherence with the world. If 1/6 + 1/2 + 1/9 = 3/18 + 9/18 + 2/18 = 14/18 = 7/9 on paper, is it also true that 1/6 of a pie plus 1/2 of a pie plus 1/9 of a pie makes altogether 7/9 of a pie? It is true. Praise the Lord for his wisdom!

14

Subtraction and Negative Numbers

Our next topic is negative numbers. We might ask, about negative numbers, the same question that we asked about fractions: are they real? Negative numbers seem to some people to be even more fishy than fractions. They ask, "Can there be a collection with a negative number of members in it? If not, aren't negative numbers a mere figment of the mind of mathematicians?" What about Kronecker's dictum, "God made the integers; all else is the work of man"? Are negative numbers the work of man?

Ledgers, Budgets, and Debts

Situations in the world illustrate the idea of counting negatively. When a family or a government is trying to balance its budget, it reckons with income and expenses. The income is "positive." The expenses are "negative." The budget is "balanced" if the income and expenses match. Even better than a balanced budget is one where there is a little surplus: the income is more than the expenses, as a cushion. The surplus in the budget is the difference between the income and the expenses, calculated by *subtraction*.

For example, suppose that the monthly income for a family is $2,000 and the total expenses are $1,900. What is the surplus at the end of the month? We rely on the fact that God has established financial regularities in this world. Money does not disappear into thin air; nor does it materialize from nowhere. The total amount of money that comes in during the month (the income of $2,000) must all go somewhere. Some of it goes out of the house to pay expenses. The rest, the surplus, will still be there at the end. So the expenses plus the surplus are equal to the income. $1,900 plus surplus is $2,000. So what is the surplus?

$100 is the right amount to complete the addition problem: $1,900 + $100 = $2,000. Situations like these are common. So schools teach children how to solve the problem by subtraction. Subtraction *undoes* addition. If $1,900 + $100 = $2,000, then $2,000 - $100 = $1,900 and $2,000 - $1,900 = $100.

We can apply the normative, situational, and existential perspectives to subtraction. First, consider the normative perspective. There are *rules* or norms for doing subtraction right. If we do not follow the rules, the money at the end of the month will not match what we calculated. Second, in the situational perspective we focus on the situation, which involves income and expenses. There is only so much money. Third, in the existential perspective we focus on the persons. In this case, the person involved in the situation is doing a calculation, either mentally or on paper. The person has to understand the meaning of subtraction, and its relation to the problem of figuring the surplus at the end of the month. He also has to know the rules for subtraction if his work is going to come out right.

As usual, we can observe that God has ordained all three perspectives. He put in place the norms; he has created the situation; and he has created the people who can think his thoughts after him. He has ordained all three in such a way that they are in harmony. The budget maker depends on the harmony in working out the budget. The point here is that, even though subtraction is conceptually more complex in some ways than addition, both addition and subtraction are due to God.

We can also see the principle of one and many. The one in this case is the general principle that 2,000 - 1,900 = 100. The many are the many cases in the world for which this arithmetical truth holds: a household budget, or a business budget, or a business inventory, or a farmer's harvest. The one and the many interlock, based on their foundation in God.

Negative Numbers

Now we can introduce negative numbers. Suppose that in one month the family income is $2,000 and the expenses are $2,100. What is the surplus? The rule says that the surplus is the difference between income and expenses, namely $2,000 - $2,100. But this is no longer exactly the same kind of subtraction problem, because the expenses are greater than the income. We say that the household has a *deficit* of $100, not a surplus. If the family has put away some savings, they can dip into the savings to tide themselves over. Let us say that they subtract $100 from a total savings of $500, leaving them with $400 in savings. On the other hand, when they have a surplus of $100 in one month, they can deposit it in the savings, and if they started with $400 in savings, they will have $400 + $100 = $500.

A surplus functions like an addition to savings or cash-on-hand. A deficit is like a subtraction. It is negative. We can also imagine the family going into debt, so that they owe $100 to a friend or to the bank or to a credit card company. The debt is also a negative amount, because it is something from which the family has to recover in order to get to a debt-free situation. They can become debt-free only by counteracting the debt with earnings.

There are many situations like this one. Such situations in the world are a justification for the concept of a negative number. A negative number is simply a number on the "other side" or the negative side of a ledger or a budget or a system for tracking quantities. In the total process of reckoning, it will be subtracted away from the total, whereas numbers on the positive side will be added in. By virtue of the commutative and associative properties of addition, it makes no difference what order we use to do the additions and subtractions. Each case with a budget or a tracking system is a particular instantiation of the principle of negative numbers.

One particular helpful illustration of negative numbers uses the number line. It is so helpful that teachers frequently use it in the classroom to teach the concept of a negative number. The number line is like a yardstick with markings on it for successive numbers, 1, 2, 3, etc. The numbers get bigger going to the right. On the left of 1 is 0, which corresponds to a balanced budget. To the left of 0 is -1, which can signify being 1 dollar in debt, or being 1 short of a required quantity. To the left of -1 is -2, then -3, and so on. In that direction one gets further in debt. The distance between two numbers on the line represents how much one would have to gain or lose to go from one position to the other.

As usual, we can apply normative, situational, and existential perspectives to this representation through a number line. The normative perspective focuses on the rules for adding and subtracting on the number line, and the coherence between three representations. We have (1) the spatial representation through a line; (2) the numerical representation through numbers on paper; and (3) the mental representation through ideas in people's minds. The situational perspective focuses on budgets and inventories and other situations in the world where there can be addition to and subtraction from a total amount. The existential perspective focuses on people's understanding of how a number line works and how budgets work.

One of the "fishy" properties of a negative number is that the negative of a negative is a positive: -(-2) = 2. This rule seems counterintuitive to many people when they first hear it. But it has an illustration in the world. If Bill is $5 in debt to Charlie, his situation is represented by the number -5. The minus sign is there because Bill is below zero by being in debt. If Bill adds $2 more to his debt, it is represented by adding -2. The negative sign indicates that the 2 is a debited 2, rather than a credit of 2. (-5) + (-2) = -7, for a total of $7 debt (the negative sign on 7 also indicates debt). But suppose Charlie tells Bill that he will forgive $2 of the $5 debt. Forgiveness is the negative of adding to the debt. It is the negation of -2, or -(-2). Bill is now only $3 in debt. -5 -(-2) = -3. The net result is the same as if he had received $2 as credit, that is, a positive 2.

An alternative explanation would say that the rule -(-2) = 2 is the only way of preserving the normal laws of addition and subtraction. Suppose that we write 6 - 3 = 3. This can also be written as 6 - (1 + 2) = 3 or 6 - 1 - 2 = 3. When we drop the parenthesis, the minus sign preceding the parenthesis has to be applied to all the numbers within the parentheses. In effect, an entry of several numbers such as 1 and 2, one after the other, on the debit side of the ledger has the same result as calculating the sum of all the debits normally (1 + 2) and then putting the resulting sum on the debit side (-(1 + 2) = -3). Now observe that 6 - (5 - 2) = 3. If we carry through the same rule about applying the minus sign, it comes out 6 - 5 -(-2) = 3, which simplifies to 1 - (-2) = 3. Clearly this will work only if -(-2) = 2.

The explanation with Bill and Charlie is oriented to the situation of debt, and uses the situational perspective to explain the rationale for -(-2) = 2. The explanation in terms of laws of addition uses the normative perspective. Both lead to the same conclusion, because God has ordained harmony in numbers.

The Nature of Negative Numbers

Negative numbers, like rational numbers, may seem to be a kind of human "invention" when we compare them to the starting point with positive whole numbers. They involve an additional effort in human understanding. That effort is part of the focus of the existential perspective. They also involve an invention in notation (using the subtraction sign "-" in a new way). This invention of notation is something that we as persons do, so it is in focus in the existential perspective. But the involvement of the other two perspectives shows that negative numbers are not *mere* invention. They correspond to norms and to situations in the world. In addition, God knew about this "invention" before human beings did. Human beings are thinking God's thoughts after him. It follows that negative numbers have a reality, in relation to the purposes that they serve in budgets, in physical measurements, and in other areas of study.

Zero

Similar observations can be made about the "invention" of zero. To have a notation for zero is an important part of the decimal system of notation, which enables us compactly to write larger numbers like 20 and 1,003. Zero has a relation to notation, to our minds (the existential perspective), to norms (2 + 0 = 2), and to situations in the world (a balanced budget has 0 surplus and 0 deficit). These perspectives harmonize according to God's plan.

15

Irrational Numbers

Next we consider irrational numbers. The name *irrational* already hints at a history in which some people had difficulty with them.[1] Most mathematicians consider them thoroughly rational, in the ordinary sense of the word, but the historical label *irrational* has stuck.

Definitions of Rational and Irrational Numbers

A *rational number* is a number that can be expressed as a ratio a/b of two integers a and b. Rational numbers include whole numbers (3, 11, 524), negative numbers (-2, -13), fractions (1/3, 2/19), improper fractions (fractions greater than 1: 12/5, 14/3), negative fractions (-1/3, -12/5), and mixed numbers (2 ½, 5 ¾). (Mixed numbers are just an alternate way of writings improper fractions: 2 ½ = 5/2.)

An *irrational number* is a number that is not rational but that still represents a quantity. The square root of 2, designated $\sqrt{2}$, is one such number. It is defined to be the number such that its square is 2; that is, when multiplied by itself the result is 2:

$$\sqrt{2} \times \sqrt{2} = (\sqrt{2})^2 = 2.$$

The square root of 3, designated $\sqrt{3}$, is another irrational number. $\sqrt{3} \times \sqrt{3} = 3$. However, the square root of 4 is rational. $\sqrt{4} = 2$, because $2 \times 2 = 4$, and 2 is rational.

How do we know whether a square root is rational or irrational? It can

[1] Actually, the history is complicated. It involves the intertwining of two meanings of Latin ratio, which can mean either "reason" or "ratio." Behind that is the Greek term *logos*, which again can mean either "reason" or "ratio."

be shown by strictly mathematical argument that the square root of any whole number is irrational, except in the case where the whole number with which we start is a perfect square, that is, when it is the square of some other whole number.

The ancient Greeks associated with the Pythagorean school discovered the difficulty with irrational numbers. The difficulty is connected to the Pythagorean theorem, which says that in any right triangle the square of the length of the hypotenuse is equal to the sum of the squares of the sides (diagram 15.1).

Diagram 15.1: Pythagorean Theorem

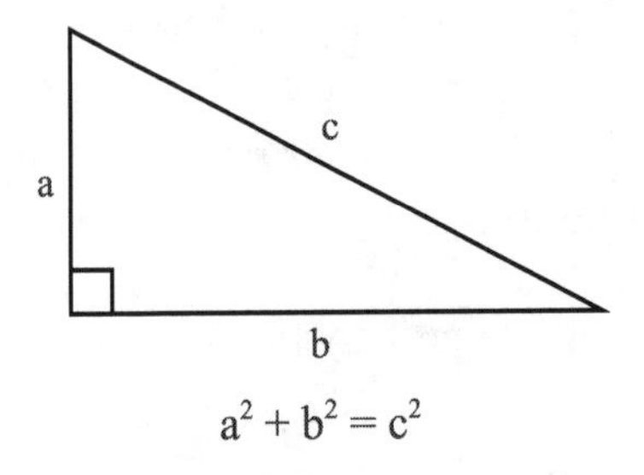

$$a^2 + b^2 = c^2$$

If the two sides have a length of 1 unit each, the square of the hypotenuse must be $1^2 + 1^2 = 2$. The hypotenuse itself must then have a length equal to the square root of 2, which is irrational. This result upset the Pythagoreans, because they had a philosophical desire to see the whole universe in terms of ratios of numbers, and the square root of 2 could not be expressed as a ratio of whole numbers.

Can we say anything coherent about the square root of 2? We can say approximately what it is, using decimals. The decimal notation is a convenient way for writing numbers in terms of powers of 10. For instance, 536 means 5 hundreds (5×100), plus 3 tens (3×10), plus 6 ones (6×1). This procedure can be extended to deal with fractions. So 1/2 is 0.5 or 5 tenths ($5 \times 1/10$). 1/4 is 0.25 or 2 tenths ($2 \times 1/10$ or 2×0.1) plus 5 hundredths ($5 \times 1/100$ or 5×0.01). 1/3 is 0.33333 . . . . The decimal representation for 1/3 does not terminate: the sequence of threes goes on forever. But 1/3 can be approximated to any desired degree of accuracy by including enough decimal places. And this is usually what is done with electronic calculators (though the internal workings of a calculator convert the decimal representation to binary representation, and then reconvert to decimals at the end of the calculation).

All fractions can be represented in decimal notation. Some of the decimals terminate (1/8 = 0.125). Others go on forever (1/6 = 0.166666 …). The ones that go on forever repeat a pattern. Thus 1/9 = 0.11111 … . The pattern may be several digits long:

1/7 = 0.142857142857142857142857142857142857142857 … .

By contrast, the decimal representation of an irrational number does not repeat:

$\sqrt{2}$ = 1.4142135623730950488 …
$\sqrt{3}$ = 1.7320508075688772935 3 …
$\sqrt{4}$ = 2.00000 (rational)
$\sqrt{5}$ = 2.2360679774997896964 1 …

It may feel as if irrational numbers are not "under control," since we cannot represent them exactly in a decimal expansion. But in a sense we cannot do that for many rational numbers like 1/3 or 1/7, since the decimal expansion goes on forever. Moreover, for square roots or cube roots or many other irrational numbers, we can program a computer to calculate the value to any desired degree of accuracy.

The concept of increasing accuracy includes within it the idea of traveling toward a limit that is never actually reached. It has an affinity to what we observed earlier about our conception of the natural numbers. We never reach the end or complete the process of listing the natural numbers. Similarly, we never reach the end of the decimal expansion of $\sqrt{2}$. Our finiteness makes it impossible actually to reach the end. Nevertheless, by an imitation of God's transcendence we can conceive of an indefinitely extended process. The irrational number is a kind of wrapping up of the entire process, once we conceive of it as a whole. Thus, we are relying on the infinity of God as the foundation for our conception of irrational numbers.

Irrationals in the World?

Are there instances or embodiments of irrational numbers in the world? We can cut up an apple into 4 pieces, each of which is 1/4 of an apple. But could we have $\sqrt{2}$ apples? It does not seem so. Because we are finite, we cannot make an infinitely sharp division, either of an apple or of a measuring stick. The hypotenuses of some right triangles offer us instances where irrational numbers crop up. But a right triangle is an idealization. Triangles that we draw on paper do not have perfectly straight lines, and the lines are not infinitely thin, and we cannot guarantee that the base angle is exactly a right angle. Even if we could guarantee all of these things, we could not guarantee that the two sides would have *exactly* the same length.[2]

Clearly, we can conceive of irrational numbers in our minds. And we can set up ways of calculating their values. In addition, irrational numbers can occur indirectly in scientific theories about the world.[3] When the scientific theories match well with experiments, we are assured that they have relevance to the world.

Nevertheless, irrational numbers do not have quite the "immediacy" of relevance to the world that we can illustrate with small fractions like 1/2 or 2/3. But consider the fraction 2,056,197,131/5,414,760,808,353. Does it have immediate relevance? If not, is it any more or less "real" than the irrational number $\sqrt{2}$?

God has made us in his image. When we try to think his thoughts after him, we can find ourselves thinking *beyond* the immediacy of our environment. We can extend our minds, and grasp the meaning of a fraction like 2,056,197,131/5,414,760,808,353 that we will never have an opportunity to use in a practical way. Likewise, we can grasp the meaning of $\sqrt{2}$.

So why were numbers like $\sqrt{2}$ called *irrational*? Perhaps one source of uneasiness lay in the feeling that a person could never *master* such a number. He could never completely control it in his mind, through a

[2] In addition, Einstein's general theory of relativity implies that a triangle in the real world is slightly "curved" by a gravitational field such as the field produced by the earth (or even the field produced by the body of an observer). So the properties of Euclidean geometry do not hold precisely for triangles in our universe.

[3] In particular, $\sqrt{2}$ occurs in quantum mechanics in connection with the principle of superposition. It does not seem to be dispensable.

direct knowledge of the entirety of its decimal representation. But if we acknowledge God as the source for *all* our knowledge, we ought to acknowledge that we can never completely master *anything at all*! There is mystery in all our knowledge, because it all reflects God who is infinite. The irrational numbers just illustrate mystery more obviously.

Perspectives

We can see the hand of God in irrational numbers. He has provided three perspectives. In the normative perspective, we have fixed rules or norms for irrational numbers. We can give rules for calculating their values as precisely as we want.[4] We can give rules for using them in calculations. In the situational perspective, we see that they have relevance to the world indirectly, through scientific theories that use them. In the existential perspective, we can see that we as human beings can understand both the rules and the applications to the world.

As usual, God's harmony with himself guarantees the harmony between the three perspectives. God has given us the fascination and mystery of irrational numbers, as one aspect of a rich world and rich minds that think about the world.

[4] There are exceptions to this kind of rule in the case of irrational numbers that can be proved to exist, but where we know of no recipe for calculating them.

16

Imaginary Numbers

Next we consider *imaginary numbers*. From our previous survey of different kinds of numbers, the pattern should be clear. As we go, we are traveling further away from the world of everyday experience. But God gives coherence to these more distant regions, just as he does to everyday experience. He knows all about these things before we do, and his own archetypal coherence provides the foundation. We can enjoy each area of mathematics as a gift from him. When we see beauty, we can thank him and praise him for it, because it reflects his original beauty.

What Are Imaginary Numbers?

The expression *imaginary number* hints at the difficulty that people found historically in trying to decide about the legitimacy of a new region of mathematics. It is indeed new, in comparison with everything that we have discussed so far. Historically, imaginary numbers were "manufactured" numbers, deliberately introduced as an "artificial" product, in order to supply solutions to equations that otherwise would have no solutions, or would not have enough solutions.[1]

Consider the equation

$$x^2 = -1$$

Can we find a solution? That is, can we find a number x whose square is

[1] The history is complex. Imaginary and complex numbers were introduced both for the solution of cubic equations and to explore quadratic equations that have no solution in real numbers (see Orlando Merino, "A Short History of Complex Numbers," January, 2006, http://www.math.uri.edu/~merino/spring06/mth562/ShortHistoryComplexNumbers2006.pdf).

-1? The square of 1 is 1. The square of -1 is also 1, since the product of two negative numbers is positive: $(-1) \times (-1) = 1$. It will not help to look for a solution among rational or irrational numbers, since every square of a positive or negative number is positive. We could say therefore that the equation "has no solution."

But mathematicians have tried to see what happens if they "invent" a solution, in a way analogous to extending the number system from whole numbers to fractions, from there to negative numbers, and from there to irrational numbers. The mathematicians simply make up a new symbol. The standard symbol is i. Mathematicians *define* i as a "number" whose square is -1. They assume that the same basic laws hold for this new "number" as for ordinary numbers. (Note that, in referring to laws, they use the normative perspective.)

Once we have i, we can form multiples of i: $2i$, $i/4$, and $i\sqrt{3}$. These also are imaginary numbers. The expression *complex number* is the name given to numbers formed by adding a real number (rational or irrational) to some multiple of i. For example, $2 + 5i$, $1 - 3i$, $1/2 + 3i/2$, and $\sqrt{3} + i\sqrt{2}$ are complex numbers. The numbers that do not involve i are called *real numbers*, to indicate that they are distinct from complex numbers.

The normal laws of arithmetic, concerning addition, subtraction, multiplication, and division, still hold for these new numbers, the complex numbers. Using complex numbers, mathematicians can provide solutions to any algebraic equation whatsoever. For example, by allowing for complex numbers, they can show that any quadratic equation $ax^2 + bx + c = 0$ has exactly two solutions (or one solution occurring twice). Some quadratic equations already have solutions using real numbers. For example, $2x^2 + 3x + 1 = 0$ has two solutions, $x = -1$ and $x = -1/2$. But other equations, like $x^2 + 1 = 0$ or $x^2 + x + 1 = 0$ have no solutions using only real numbers. Likewise, if we allow complex numbers, any cubic equation $ax^3 + bx^2 + cx + d = 0$ has exactly three solutions (or one solution occurring three times, or two solutions, one of which occurs twice). On the other hand, if we refuse to use complex numbers, quadratic equations may or may not have any solutions. Complex numbers have won the hearts of mathematicians, not only because of this beautiful result, but because of many other beauties in the theory of complex functions.

Complex numbers have also won the hearts of scientists through applications to the world of science. Particularly notable is quantum mechanics, which uses complex numbers in an indispensable way. Why should it be that this "invention" out of the minds of mathematicians, who were looking for beauty in the world of abstract mathematics, should centuries later find applications in physics? God in his wisdom has made it so.

Perspectives on Imaginary Numbers

As usual, we can consider imaginary numbers from the normative, situational, and existential perspectives. The normative perspective observes that imaginary numbers and complex numbers obey the same basic laws as ordinary numbers, and they behave consistently. The beautiful properties of these numbers come from God, who is beautiful. That is some assurance that these numbers are "real," because they are known by God, rather than merely "imaginary."

The situational perspective observes the applications of imaginary numbers to the world of science. There is coherence between the normative laws and the way the world works.

The existential perspective observes that we can coherently understand and reason about these numbers. Our reasoning, when done right (normatively!) is in harmony both with the objective norms from God and with the world.

17

Infinity

Is infinity a number? We have already met with the idea of infinity in connection with the natural number system. The list of natural numbers, 1, 2, 3, … , extends indefinitely. We might say that it goes on "forever" or that there is an *infinite* number of natural numbers. In the appendix on set theory (appendix E) we consider infinite sets, some of which are in a sense even "larger" than the set of nonnegative integers. How should we regard the idea of infinity?

Human Limits

The idea of infinity leads straight back to the earlier discussion that we had about human knowledge in comparison to God's knowledge (chapter 5). We are finite. At the same time, we know the infinite God. We can know about infinity by knowing God. Our knowledge is genuine, just as our knowledge about God is genuine, without being exhaustive. We do not *comprehend* God, in the special sense of the word *comprehend*. We do not ever achieve mastery in our knowledge of him, nor do we know everything that he knows, nor do we know it in the same *way* that he does. There is mystery for us, because God exceeds our grasp.

The same truths are relevant when we consider infinity in mathematics. As finite human beings, we never come to the end of the sequence of natural numbers. We never exhaust infinity. Rather, by imitating God's transcendence on our finite level, we see the general pattern of progression in the number sequence, and we imagine its indefinite extension. The same kind of principle applies to all the cases where the idea of infinity crops up in mathematics. Infinity is a kind of limit or extrapolation from our mind, and we know well enough, when we reflect about our knowledge, that we never literally attain it. Nevertheless, we can work with the idea, because God has given us the ability to do so, as people made in his image.

Modern set theory (appendix E) has given us not one infinity but many. Besides the set of natural numbers, there are additional sets, such as the power set of the set of natural numbers, and power sets built on top of that, that extend upward indefinitely. Set theory can define when two sets represent the "same level" of infinity, namely when we can establish a one-to-one correspondence between the members of the two sets. But it can also be shown that there are "bigger" sets that are not in one-to-one correspondence with the natural numbers. The details must be left to the technicalities of set theory. But the result is that there is a whole series of bigger and bigger infinite sets.

What do we do with these ideas? Some mathematicians, the "finitists," are suspicious of all infinities, even the smallest one, the infinity of natural numbers. Others embrace the whole series of larger infinities with delight. I am closer to the latter group, because I see the beauty of God's archetypal infinity reflected in the towers of infinities in set theory. God is good, and he has given us many wonders. The wonders include not only the beauty of mountains and flowers and sunsets, but—for those who have the ability to appreciate them—the beauties of mathematics and the beauties of these infinities in set theory. It is all due to him. We can embrace these infinities as a gift, and rejoice.

At the same time, I have a sympathy for the finitists. They have a point—a grain of truth in their favor. They are rightly sensitive to the issue of human limitations and human finiteness. They rightly understand that no one who is a human being can *comprehend* or *exhaust* infinity. The sets that are called "infinite" sets are manipulated by mathematicians because we have symbols and rules for manipulation. The rules represent truths about miniature transcendence, and its imitation of God's transcendence. But they do not literally create infinite sets as objects in the world, containing, let us say, an infinite number of atoms. (Only a finite number of atoms exist within the visible universe.)

When we reason about infinite sets, we are continually projecting beyond our limitations, on the basis of the analogy between our minds and the mind of God. We do not fully understand what we are doing. And indeed, in the early days of the theory of infinite sets, as developed by Georg Cantor, investigators confronted paradoxes.[1] It is easy to produce a contradiction if we let our reasoning run away with us and do not exercise restraint. The history of set theory in the twentieth century can be understood largely as a history of exploring and wrestling with our limitations, and how we can avoid contradictions in our reasoning while we are stretching our reasoning into realms that we can never fully master.

We best explore this realm when we do it for the glory of God and for his praise.

[1] See Poythress, *Logic*, appendices A1 and E2.

Part V

Geometry and Higher Mathematics

18

Space and Geometry

Now let us turn to consider geometry. How does it relate to God? We may confidently assume that it is due to him, but can we say more?

Space

Geometry as a subdiscipline within mathematics receives some of its motivation from our ordinary experience of space. The ancient Egyptians and Babylonians had to work out spatial relationships and measurements with care in preparation for their great building projects, and geometrical observations of a practical kind can be found as early as 1900 BC. Geometry seems to have had a practical origin in connection with measurements in space. It reached a classical rigorous formulation in Euclid's *Elements*.

So let us start with the idea of space. The description of creation in Genesis 1 implies that God has ordained the spatial structure of the world in which we live. For the most part Genesis 1 focuses on things and actions within space, rather than on space itself. But it does indicate that God "separated" major regions. God separated the heaven from the earth in Genesis 1:6–8, and the sea from the dry land in verses 9–10. In Genesis 1 as a whole God created a large-scale dwelling place, which is filled with his presence (Jer. 23:24).

The tabernacle of Moses, as we observed (chapters 7–8), is a miniature dwelling place, a "copy" or "image" of God's larger dwelling place in heaven and in the universe as a whole. The tabernacle exhibits simple numerical relationships and numerical proportions. At the same time, it exhibits spatial relationships and spatial proportions in the two rooms and in some of the items of furniture (the table for the bread of the presence is 1 by 2 by 1 ½ cubits). So the tabernacle invites us to see a relationship between its shapes and the "shape" of the larger world, including its spatial characteristics.

We can ask about the archetype for the tabernacle. The tabernacle rooms are images or shadows of God's heavenly dwelling place among the angels. And does this dwelling place have a deeper root? It does. The tabernacle and heaven both point forward to Christ, who is the dwelling place of God (John 1:14; 2:21). The New Testament indicates further that Christ's fellowship with God the Father existed before the world began (John 1:1). This fellowship takes the form of *indwelling*. John 17:21 indicates that the Father is *in* the Son and the Son is *in* the Father, in a context that reflects back on eternal Trinitarian relationships (17:5, 24).

This mutual indwelling, which includes the Holy Spirit, is called *coinherence*. Since God is unique and infinite, the indwelling of the persons of the Trinity in one another is mysterious to us. It is *not* a spatial relationship in the way that we experience it in the created world. God is not spatially divisible, as if one part of him could be here and another there. It is not "spatial" at all, if what we mean by "space" is determined just by our experience of space in the created world.

Does that mean that the language of indwelling means nothing at all? No, it does have meaning to us. It is indicating that there is an *analogy*, but not identity, between indwelling in the tabernacle, or the Holy Spirit dwelling in a believer (John 17:23; compare 14:16–17, 23), and the archetypal indwelling among the persons of the Trinity. The archetype, as usual, is not equal to the ectype. Nevertheless, there is a relationship between the two, as indicated by the expressions used. We cannot *comprehend* this relationship, because we are finite and God is God. But we can understand the reality of analogy between the Holy Spirit dwelling in us and the Old Testament symbol of the temple: "Or do you not know that your body is a *temple* of the Holy Spirit *within* you, whom you have from God?" (1 Cor. 6:19). The temple in turn, as we have observed, evokes analogies with the tabernacle of Moses, with God's heavenly dwelling, and then finally with the archetype, the coinherence of persons in the Trinity.

We may take the analogy one step further, and move from the picture of the temple to the universe as a whole. The universe is the large-scale dwelling place of God. So the spatial character of the universe has its archetype in God, and more specifically has an archetype in the coinherence of persons in the Trinity.

Laws of Space: Geometry

Space reflects God's presence, and so testifies to its creator. And laws concerning space, such as the laws of geometry, have their origin, like all laws, in the speech of God.

But what kind of space are we talking about? Here we must see that Euclidean geometry, such as was axiomatized by Euclid and later refined by mathematicians like David Hilbert, is related to space as we experience it, but is an idealization. If we draw a line on paper, it is not perfectly straight, even if we use a ruler to help us. It is also not perfectly thin (no width). Its intersection with a second line is not a dimensionless point, but a bit of ink or a bit of pencil graphite that covers a small area. The idea of a dimensionless point and the idea of a line with no width are extrapolations, for the sake of avoiding the distractions and complexities involved with lines that are 0.2 mm wide.

Euclid's geometry illustrates the interlocking of the one and the many. Consider a particular theorem within Euclid's geometry, namely the theorem that the two base angles of an isosceles triangle are equal. (An isosceles triangle is a triangle in which two of the sides are of equal magnitude. The two angles opposite these two sides are then also of equal magnitude.) This theorem is a *general* theorem. We are to understand that it holds for all the particular cases of isosceles triangles, of various sizes and shapes. The particular cases are many. The one truth is one. We understand the meaning of the one truth through its many illustrations, and likewise we understand the full meaning of the property of equal angles in a particular triangle when we see it in relation to the general theorem. The general theorem makes it possible for us not to repeat our reasoning every time we have a new instance of an isosceles triangle. As we noted before (chapter 2), this interlocking of one and many depends on God.

In the twentieth century the situation has turned out to be even more complicated. Albert Einstein's general theory of relativity postulated that space (together with time, which is treated as a fourth dimension not strictly isolatable from the spatial dimensions) is curved, not Euclidean. Euclid's famous parallel postulate turns out not to be strictly true of the space in which we actually live.[1] The discovery of non-Euclidean geometries (where the parallel postulate did not hold true) shocked the intuitions of many mathematicians, and the physicists were even more shocked when they heard from Einstein that these non-Euclidean geometries had relevance to the real world.

Perspectives on Space and Geometry

What should a Christian think? We are seeing here a complex relationship between John Frame's three perspectives. Human intuitions are in focus for the existential perspective. The intuitions, until the work on non-Euclidean postulates in the nineteenth century, said that space had to be Euclidean. But of course the intuitions had been trained by hundreds of years of dominance by Euclid's *Elements*, the classical text on geometry. If people had paused to notice, they could have seen all along that Euclid presented an idealization and that Euclid's theorems exhibited the mystery of the interlocking of one and many. These characteristics could make it easier to admit that God may do as he wishes, and that the world we live in might not be Euclidean.

Frame's normative perspective focuses on the laws of geometry. But the laws that Euclid formulated are an idealization. So they approximate but do not necessarily match what God has ordained to be true for the world. This approximation, of course, was in the mind of God before it was in our minds. God had it prepared as a stepping stone in the process by which human beings would grow in understanding God's world and grow in praising him. Euclid's formulation is still useful as an axiomatic

[1] See Poythress, *Logic*, chapter 54.

system for the world of the mind, and it is used today by physicists and mathematicians who are well aware that it does not perfectly match the world around us.

Finally, we consider the situational perspective. This perspective focuses on the world. Here is where we appreciate the world as God has given it to us, and we consent to believe that it is non-Euclidean, according to Einstein's description. We should also recognize that Einstein's description is not ultimate either. It is an insightful idealization of some aspects of the world, not all. If we take to heart the fact that God made a world of great richness, we avoid the temptation to be reductionistic about space and geometry, as well as other fields.

Analytic Geometry

Another question confronts us, namely the relationship of space to numbers. There is a rich relationship. René Descartes invented analytic geometry, which was a rigorous way of describing lines and shapes in space using algebraic, numerical tools. To each point in two-dimensional space is assigned a pair of numbers (x, y), where x is the distance of the point from a fixed vertical axis and y is the distance of the point from a fixed horizontal axis (diagram 18.1).

Diagram 18.1: X and Y Axes

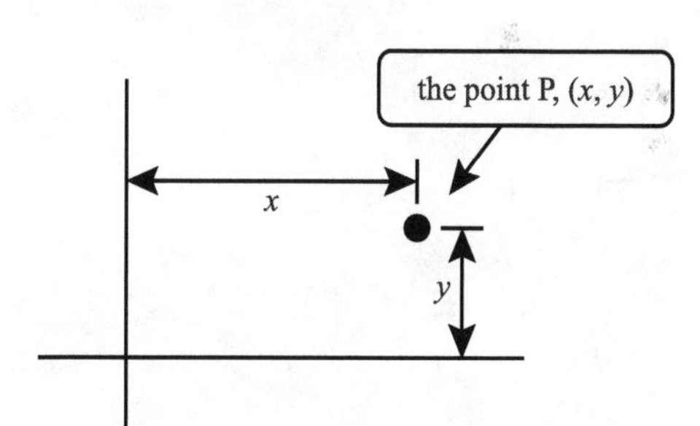

This technical arrangement allows people to use numbers not only to talk about a single point, but about a straight line, a circle, an ellipse, a parabola, and other geometrical objects. The arrangement uncovered many beautiful harmonies between the two realms, the realm of number and the realm of space.

You can imagine that these harmonies tempt some people to try to *reduce* space to number. Space, in their thinking, is just number in another form. But our ordinary experience contradicts this claim. If we take God into account, we can infer that God gives us ordinary experience, and not *just* the later mathematical analysis, as one aspect or form of reality. So the reductionistic philosophical attempt is not justified. Rather, we should say that God ordains harmony between the two realms. The harmony is so thorough that the properties of one can be deduced from the other. The relationships do not simply go one way, from number to geometry. It is also possible to represent numerical truths in geometrical form. For example, the addition of two numbers can be represented in space by using a number line. We have one line segment, of length 2 to represent the number 2, and another line segment, of length 3 to represent the number 3. When we lay them head to tail, the total length is 5. (See diagram 18.2.)

Diagram 18.2: Addition within a Coordinate System

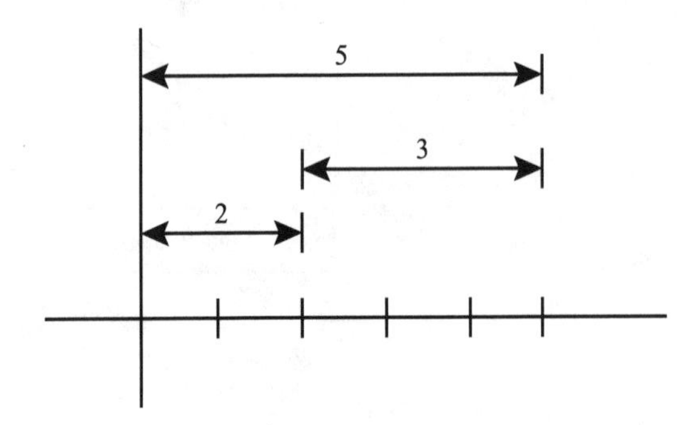

Real Numbers

One of the questions that arise when we look at space is the question of continuity. We can picture ourselves moving along a line gradually, until we arrive at our destination. The gradual motion is a continuous motion, without jerks or pauses. This perception leads by extrapolation to the idea of space being infinitely divisible and smooth in between any divisions, as minute as they might be. The points along a line are comparable to real numbers, expressed in infinitely long decimal expansions. Both infinities, the one in space and the one in time, are idealizations. As usual, we can recognize our finiteness. But we can also recognize that we have ability, given by God, to explore these idealizations and see what happens. From these sources comes the theory of real variables.

And a beautiful theory it is. It builds on intuitions coming both from our experience of number and our experience of space. But it also travels beyond them. And mathematicians try to make "rigorous" the ways in which they travel beyond them. In the development of the theory of real variables, as in the development of set theory, paradoxes were encountered when the human mind tried to push toward infinity. The story of these developments is best left to other books. The paradoxes once more indicate the mystery associated with our being finite and our also being able to think God's thoughts after him.

19

Higher Mathematics

Over the centuries mathematicians have developed more and more subdisciplines, and have continued to uncover extraordinary as well as ordinary beauties in new results. These newer areas of exploration are all gifts of God, and all reflect the beauty, wisdom, and faithfulness of God. They all become motivations for praise for those who have come to love God through the work of Christ.

Subdisciplines

The new subdisciplines have arisen mostly through processes that involve recognition of common patterns and common structures belonging to more than one instance within already existing branches of mathematics. For example, elementary algebra builds on arithmetic by seeing common patterns belonging to many instances in which we deal with numbers. Abstract algebra builds on elementary algebra, and generalizes from patterns seen in common algebraic operations such as addition and multiplication.

In the discernment of common patterns, we see reliance on the interlocking of the one and the many. The one in this case is the common pattern. The many are the instances that illustrate or display the pattern. Abstraction, which is a common feature of mathematics, is a process of focusing on the one, the common pattern, in the midst of the many.

To some extent mathematics has also been influenced by mathematical problems posed within physics and other sciences. In the relationship between mathematics and the sciences we see a confirmation of the harmony between disciplines, a harmony that goes back to God, who ordained them all.

The Discrete and the Continuous

A major distinction of subfields within mathematics arises from the difference between structures that are *discrete* and structures that are *continuous.*[1] Roughly speaking, a discrete structure is one in which every individual element is isolated from every other. A continuous structure is one in which individual elements belong to a whole in which one can move continuously from one element to another. No element is isolated. Discrete structures have a close relation to whole numbers, which are the most intuitively accessible instance of a discrete structure. Each number is distinct from its neighbors. Continuous structures have a close relation to space and geometry. Our intuitive starting point for understanding the idea of continuity uses pictures from space.

Algebra, in the most general sense, is the study of discrete structures. *Geometry* and *topology*, which is a kind of generalization of ordinary geometry, study continuous structures. But the two sides enrich one another through interaction. Algebraic geometry and algebraic topology show by their names that they are combination disciplines. Real and complex analysis use the idea of continuity in a vital way, but still deal with elements that are number-like. So these disciplines display the fruitfulness of crossover. Analytic number theory uses real and complex analysis on the way to answering questions about the natural numbers. It too involves an interaction of the continuous ("analysis") and the discrete (natural numbers).

Reliance on God

All these disciplines rely on a starting point that involves our intuitions about number, or our intuitions about space, or both together. The disciplines also show an interaction of normative, situational, and existential

[1] See Willem Kuyk, *Complementarity in Mathematics: A First Introduction to the Foundations of Mathematics and Its History* (Dordrecht-Holland/Boston: Reidel, 1977).

perspectives. The normative perspective is the most common to use in exposition of mathematics, because textbooks and explanations focus on establishing truths about mathematics. At the same time, problems and exercises show how to apply the truths to particular examples. And many areas of higher mathematics have applications in the sciences, so that the situational perspective is appropriate. When we observe that mathematics depends on our intuitions about number and space, and on our ability to abstract and generalize from particular examples, we are focusing on the capabilities of human beings, and so we are using the existential perspective.

The crossover disciplines also rely repeatedly on the intrinsic harmony between number and space. As we have observed, this harmony goes back to God, who ordained them both. The distinction between number and space is the most outstanding distinction that is linked by God-ordained harmony. But in a broader sense we can see less striking distinctions throughout mathematics. Each number is what it is, and has distinct properties.

For example, the number two is even, while three is odd. Two is the only even prime (a prime is a number that has no positive integer divisors except one and itself). Three is the lowest nontrivial triangular number. (A triangular number is a number that is the sum of successive integers, beginning with 1. Three is a triangular number, because $3 = 1 + 2$. The next triangular number after 3 is $6 = 1 + 2 + 3$.) Each number has some properties that are unique to it.

In addition, each *kind* of number, such as fractions or negative numbers, has its own distinctiveness. According to our antireductionistic stance (chapter 4), each thing is what it is and is not exhaustively reducible to anything else.[2] An antireductionistic approach should positively appreciate each mathematical object as well as each carnation, each squirrel, each oak tree, and each person. We affirm the many, not simply the one, as objects of appreciation. As our appreciation increases, our praise to God should also increase. God's wisdom, infinity, and beauty are reflected in the things he has made and in the minds that he has made.

[2] For the broader context for antireductionism, see Poythress, *Redeeming Philosophy*; Poythress, *Symphonic Theology: The Validity of Multiple Perspectives in Theology* (reprint; Phillipsburg, NJ: Presbyterian & Reformed, 2001).

Conclusion

When God created the world, he also ordained all the characteristics of the world. It is he who specifies all the truths about the world, including the truths of mathematics.

God's speech reveals his character. His speech is divine, with divine characteristics, and this speech includes truths about mathematics. Because every aspect of this world reveals God's character, it is a delicate question as to what reflects the *necessity* of God's character and who he is, and on the other hand what reflects the *contingency* of the decisions God made to create a world such as the one we enjoy.

In any case, the world reflects the character of God and reveals God, so that we should respond in worship and praise. Christ the Lord is not only the creator of the world, but also its redeemer. Through faith in him we may be reconciled to God and turn from suppressing the truth about God that he reveals in the world. But the process of recovery is gradual. The Bible describes one aspect of the process as renewal of the mind:

> I appeal to you therefore, brothers, by the mercies of God, to present your bodies as a living sacrifice, holy and acceptable to God, which is your spiritual worship. Do not be *conformed* to this world, but be *transformed* by the *renewal of your mind*, that by testing you may discern what is the will of God, what is good and acceptable and perfect. (Rom. 12:1–2)

As part of the renewal of our minds, we need to be renewed in our thinking about mathematics. We need to grow in seeing it as a gift from God that reflects the giver—and to give thanks with increasing devotion. May this book help in the process.

Supplements

Resources

This book represents a beginning, rather than an end. Other people have written already about the bearing of Christian faith on mathematics, and still others will write more in the future. One major resource is found in James Nickel, *Mathematics: Is God Silent?*,[1] which contains not only much historical information but illustrations that are useful for teaching mathematics. There is also an early article, Vern S. Poythress, "A Biblical View of Mathematics."[2]

Resources for teaching and for further discussion can also be found with the Association for Christians in the Mathematical Sciences. This Association provides a forum for discussions related to Christian faith, theology, mathematics, and related fields such as computer science. It has a number of resources, including a website, http://www.acmsonline.org/, a journal (*Journal of the ACMS*), and a biennial conference (in odd-numbered years).

Steve Bishop has compiled an online bibliography for books and articles on Christianity and mathematics.[3] It references a more extensive, older (1983) bibliography by Gene B. Chase and Calvin Jongsma.[4]

I am grateful for two writings[5] of D. H. Th. Vollenhoven that originally

[1] James Nickel, *Mathematics: Is God Silent?* rev. ed. (Vallecito, CA: Ross, 2001).

[2] Vern S. Poythress, "A Biblical View of Mathematics," in Gary North, ed., *Foundations of Christian Scholarship: Essays in the Van Til Perspective* (Vallecito, CA: Ross, 1976), 158–188; http://www.frame-poythress.org/a-biblical-view-of-mathematics/, accessed December 29, 2012.

[3] Steve Bishop, "A Bibliography for a Christian Approach to Mathematics" (June 7, 2008); http://www.scribd.com/doc/3268416/A-bibliography-for-a-Christian-approach-to-mathematics, accessed September 17, 2012.

[4] Gene Chase and Calvin Jongsma, "Bibliography of Christianity and Mathematics, 1st edition 1983"; http://www.asa3.org/ASA/topics/Mathematics/1983Bibliography.html, accessed July 30, 2012. This bibliography was published by Dordt College Press in 1983, but is now out of print.

[5] D. H. Th. Vollenhoven, "Problemen en richtingen in de wijsbegeerte der wiskunde" [Problems and Directions in the Philosophy of Mathematics], *Philosophia Reformata* 1 (1936): 162–187; D. H. Th. Vollenhoven, *De wijsbegeerte der wiskunde van theïstisch standpunt* [The Philosophy of Mathematics from a Theistic Standpoint]

drew my attention to the issues in understanding mathematics from a Christian standpoint.

Those who want to set Christian thinking about mathematics within a larger context might consult some of my books that consider a larger context: science (*Redeeming Science*), probability (*Chance and the Sovereignty of God*), philosophy (*Redeeming Philosophy*), and worldviews (*Inerrancy and Worldview*).[6]

(Amsterdam: Van Soest, 1918). On the philosophy of the law-idea, see Poythress, *Redeeming Philosophy*, appendix A.

[6] Poythress, *Redeeming Science*; Poythress, *Chance and the Sovereignty of God: A God-Centered Approach to Probability and Random Events* (Wheaton, IL: Crossway, 2014); Poythress, *Redeeming Philosophy*; Poythress, *Inerrancy and Worldview*.

Appendix A

Secular Theories about the Foundations of Mathematics

People concerned with the philosophy of mathematics have discussed for a long time what mathematics really is, and what numbers are. In most of this discussion, God has been absent from the picture. And that creates difficulties. The major difficulty is that omitting God falsifies the picture not only for mathematics but for anything at all that we want to study. God is omnipresent, all-present. He is present in the whole world and every aspect of the world. In addition, he is sovereign ruler over the world, so that everything owes its existence to him. Leaving him out means leaving out the primary source both for existence and for meaning.

Reductionisms

People have nevertheless done it. In the area of philosophy of mathematics, it has resulted in reductionism. Mathematics gets "reduced" to some aspect of the world. The history of philosophy has seen several main competitors for explaining mathematics: Platonism, empiricism, logicism, intuitionism, formalism, and predicativism.[1] Each of these has a favorite starting point. This starting point becomes the preferred platform for explaining everything else in mathematics.

[1] Leon Horsten, "Philosophy of Mathematics," *The Stanford Encyclopedia of Philosophy (Spring 2014 Edition)*, ed. Edward N. Zalta, http://plato.stanford.edu/archives/spr2014/entries/philosophy-mathematics/, accessed June 18, 2014.

Platonism starts with an ideal realm, which contains abstract ideas, including numbers and all mathematical truths. This abstract realm allegedly exists before the mathematician starts his work. Empiricism starts with sense experience, such as the experience of seeing four apples. Logicism starts with logic. Intuitionism starts with human subjectivity, and especially mental intuitions about numbers and mathematical objects. Formalism starts with language, especially the polished formal languages used in mathematical proof theory. Predicativism, a view in some ways intermediate between Platonism and intuitionism, accepts the whole numbers 1, 2, 3, … as unproblematic. They can be accepted either because of their Platonic existence or through intuition. Predicativism accepts more complex mathematical objects only when these objects can be "built up" gradually from the natural numbers, by one or more stages.

According to Platonism, mathematics really derives from the ideal realm of truth. According to empiricism, mathematics arises from human generalizations, beginning from sense experience of countable objects and spatially extended objects. According to logicism, mathematics derives from logic. And so on for the other views.

We can see that each of these approaches reduces mathematics to its favorite starting point. These approaches hope to explain the whole of mathematics as an unproblematic derivation from this focal starting point. But some of the approaches, it turns out, have gaps in their explanations that cannot be filled. Others have implausibilities. And all of them suffer from a failure to explain fully the multidimensional character of mathematics that we experience *in practice* as we use mathematics in relation to the world and in relation to a variety of realms of thoughts. God ordained a world with diversity. Even though the world exhibits harmonies between physics and mathematics, between counting and space, and so on, these harmonies do not dissolve the richness.

Some of the philosophical approaches can be classified using Frame's three perspectives on ethics. Platonism and logicism have an affinity to the normative perspective. They postulate norms for mathematics, norms that stem either from an ideal realm (Platonism) or from logic (logicism). Empiricism in mathematics has an affinity to the situational perspective. It focuses on the ties between mathematics and the world that we experience. Intuitionism has an affinity with the existential perspective. It starts with the human mind, and focuses on its subjective intuitions about numbers. Formalism and predicativism are more difficult to classify. Formalism has a focus on formal languages, which have features that are normative (rules for derivations) and features that are external to human subjectivity (written language tokens are "in the world," in the situation). Predicativism believes both in norms—the objective existence of less complex mathematical objects—and the necessity of care to make sure of our intuitions with less complex objects before we build more complex ones.

Let us now briefly consider some of the difficulties involved with the various secular approaches.

Platonism

Platonism says that numbers and mathematics belong to a realm of abstract ideas, a realm that exists before mathematicians begin to study it. In their everyday work, most mathematicians tend to operate with assumptions resembling Platonism. They assume that the objects that they study exist and that the truth about them is "out there" to be obtained. In addition, from a Christian point of view we can say that Platonism is close to the truth. God already knows about all mathematical objects and all mathematical truths before human beings start their investigations. So the objects and the truths are "out there," namely in the mind of God.

Platonism originated with the Greek philosopher Plato, who maintained that genuine human knowledge is knowledge of abstract forms or ideas—the idea of the good, the idea of justice, the idea of beauty, and so on. The natural numbers, the truths about numbers, and the truths about geometry can easily be added to the list of ideas. Plato conceived of the ideas as abstract concepts that exist independently of everything else.

In addition to Plato's original view, there are possible modifications. Christian thinkers who wanted to adopt Plato altered his view by placing the ideas in the mind of God. We will discuss this Christianized view in the following appendix.

Secular Platonism for mathematics began to get into trouble in the late nineteenth century, when logicians discovered logical paradoxes that shook confidence in human beings' ability to access an alleged Platonic realm. The paradoxes included Russell's paradox about the set consisting of all sets that do not contain themselves. Does this set contain itself? Answering either yes or no leads to a contradiction. People also encountered the paradox of "the set of all sets." Does this set include itself? Then it has to be bigger than itself.[2]

If mathematicians could through their intuition directly access the Platonic realm of mathematical ideals, why did they themselves sometimes produce contradictions in the form of these paradoxes? And if they could not access the Platonic realm, what good did it do to postulate its existence?

A Christian has a different kind of answer. We distinguish between our own knowledge and intuition on the one hand, and God's knowledge on the other hand. Our own stumbling over paradoxes just indicates the limitations of finite human knowledge. It is not a failure of God, who remains consistent with himself.

In addition to these problems, Platonism does not really explain why there is a harmony between three perspectives on mathematics. The normative perspective focuses on the realm of abstract ideas; the situational perspective focuses on the usefulness of mathematics in the world around us; and the existential perspective focuses on our ability to understand mathematics and our intuitive understanding of basic mathematical concepts like the concept of number. Why do these three agree? Platonism has to introduce another factor: Plato postulated a creator (the "demiurge") to make things in the world after the model of the original abstract ideas.

Empiricism

A second philosophical approach is empiricism. Empiricism in mathematics endeavors to derive mathematics by starting with sense experience. Empiricism has fared the worst in the twentieth century. It may seem plausible to begin with ordinary experience of seeing objects in the

[2] Poythress, *Logic*, appendix A1.

world. But we already confront the problem of the one and the many. Earlier we explained how our minds can generalize from experiences of two apples to the number two, which is an abstraction in comparison to the two apples. The number two is the "one," in relation to the *many* particular instances of two apples and two peaches. This process of generalization relies on the relation of the one to the many. So empiricism also relies on this relation, which it cannot explain.

In addition, advanced mathematics has applications in physics, and it seems impossible to explain this applicability by starting merely with the simple facts about two apples or four apples. The growth of intense mathematical applications in the twentieth century has decreased the plausibility of empiricism.

A Christian answer is different. God made a world that conforms to arithmetical and geometrical laws. It is natural for us as embodied creatures to start from experiences of this world. But when we start, we start with minds made in the image of God. So our minds are in tune with the world. And we can generalize from our experiences, because our minds are also in tune with the mind of God. We can see the coherent functioning of Frame's three perspectives. In the normative perspective, we observe that our minds are in tune with the mind of God. In the situational perspective, we observe that our minds are in tune with the world, which God made. And in the existential perspective, we observe that it is *our* minds, minds of people made in the image of God, that do the mathematics.

Logicism

Logicism is associated with the work of Alfred North Whitehead and Bertrand Russell. They jointly undertook to write the three-volume work *Principia Mathematica*,[3] in which they hoped to start with purely logical principles and derive all of mathematics from these principles. But they had to include an "axiom of infinity," which postulated the existence of an infinite number of objects. This principle did not appear to be simply a matter of logic. In addition, in 1930 the program hit the rocks. Kurt Gödel

[3] Alfred North Whitehead and Bertrand Russell, *Principia Mathematica*, 2nd ed., 3 vols. (Cambridge: Cambridge University Press, 1927).

showed that no specific list of axioms could capture all the mathematical truths about whole numbers.[4] Mathematics could not be derived from logic alone.[5]

Intuitionism

Intuitionism was a route in the philosophy of mathematics started by L. E. J. Brouwer.[6] According to intuitionism, mathematics is the creation of the human mind. The focus on the human mind is akin to Frame's existential perspective. But this perspective is forced to function within a non-Christian context, as is evident from the fact that the human mind, not the divine mind, becomes the standard for truth. As we might expect, there are difficulties.

According to Brouwer, no mathematical statement ought to be regarded as either true or false until it has been proved or refuted. Intuitionism is most famous because it denies the law of excluded middle, that is, that a proposition must be either true or false.

How could anyone deny the law of excluded middle? Brouwer was concerned about mathematical propositions whose truth is at present still unknown. One such proposition, called Goldbach's conjecture, says that every even number greater than 2 is the sum of two primes.[7] As of 2014, no one knows whether Goldbach's conjecture is true. No counterexample has been found, but neither has anyone proved that it is true for *every* even number greater than 2. According to Brouwer's view, Goldbach's conjecture should be regarded as neither true nor false until we find a counterexample or a proof.

Intuitionist assumptions have led to fruitful explorations in logic. Even without accepting Brouwer's metaphysical convictions about math-

---

[4] Poythress, *Logic*, chapter 56 and appendix D1.

[5] 1983 saw the birth of *neo-logicism*, which evaded Gödel's strictures by using more powerful assumptions and more powerful logic (Horsten, "Philosophy of Mathematics," §2.1). But critics have complained that its foundational assumptions implicitly added arithmetic to logic, so there was no genuine "reduction" to logic.

[6] Poythress, *Logic*, chapter 64; Mark van Atten, "Luitzen Egbertus Jan Brouwer," *The Stanford Encyclopedia of Philosophy (Summer 2011 Edition)*, ed. Edward N. Zalta, http://plato.stanford.edu/archives/sum2011/entries/brouwer/, accessed June 18, 2014; Rosalie Iemhoff, "Intuitionism in the Philosophy of Mathematics," *The Stanford Encyclopedia of Philosophy (Spring 2014 Edition)*, ed. Edward N. Zalta, http://plato.stanford.edu/archives/spr2014/entries/intuitionism/, accessed June 18, 2014.

[7] A *prime* number is a whole number whose only divisors are 1 and the prime itself. Thus 3 is a prime, because its only divisors are 1 and 3. 5 and 7 are also primes. 4 is not, because 4 has 2 as a divisor. 9 is not, because 9 has 3 as a divisor.

ematics, a logician can explore what can and cannot be deduced once we avoid utilizing the law of excluded middle as a given assumption. Intuitionism also introduced the fruitful idea of a *constructive proof*. Roughly speaking, a proof that is *not* constructive shows that some postulated mathematical object exists, but does not actually "construct" the object or pick it out. Rather, the proof proceeds by showing that the nonexistence of such an object would lead to a contradiction. Classical mathematics accepts both constructive and nonconstructive proofs, while intuitionism accepts only constructive proofs. All mathematicians agree that there is a difference, and that the difference is logically and mathematically interesting. So all mathematicians can in principle feel free to study and search for constructive proofs. The quarrel is over the metaphysical status of nonconstructive proofs.

Thus, intuitionism has led to some fruitful mathematical ideas. But it has difficulties as a philosophy.

First, an intuitionism of a Brouwerian sort does not provide an adequate mathematical foundation for parts of mathematical analysis that are regularly used in science. It does not really explain this kind of applicability, nor does it even provide endorsement for using the mathematics in the way that scientists use it. Scientists accordingly pay no attention to intuitionism. And even most mathematicians want the benefit from regions of mathematics that cannot be established using intuitionistic principles.

Second, the key intuitionistic claim that a proposition is neither true nor false until it is proved or refuted is counterintuitive, and seems to many people to confuse truth with proof. Truth concerns what is the case. Proof concerns what human beings *can prove* or demonstrate to be the case.

As an example, consider Fermat's last theorem. This famous theorem was conjectured to be true by Pierre de Fermat in 1637, but was proved only in 1994 by Andrew Wiles. Does Brouwer's intuitionism say that it was neither true nor false until it was proved in 1994? This way of putting it seems to redefine the normal meaning of "true" and "false." Surely, according to our ordinary way of speaking, the theorem was true in 1637, but no human being *knew* it was true until Wiles produced a proof in 1994. Even then the rest of the world did not know it was true until Wiles published the proof in 1995.

Intuitionism has tried to evade this difficulty by speaking of an "ideal mathematician." The ideal mathematician can run ahead of what we know today, but still can never complete an infinite process. The difficulty here is that our limited knowledge does not permit us to say what the ideal mathematician might achieve. It leaves us with a situation where some mathematical propositions are true and known to be so, others are true and not yet known to be so, and still others are not knowable by human beings. Only the third category is viewed as neither true nor false.

There is a grain of truth to intuitionism. If no human being knows whether a particular mathematical proposition is true, do we even know for sure whether we have a clear idea of what it would mean for it to be true? Intuitionism is wrestling with the problems raised by the limitations of human knowledge and the finiteness of the human mind. Unfortunately, as a philosophy it does not bring into the picture the infinity of God's mind. It seems to assume that human minds are ultimate determiners of truth, rather than imitators of truth that God already knows. With this assumption, it concludes that a proposition cannot be true unless some human being could come to know that it is.

Formalism

The philosophy of formalism says that mathematics is the study of formal languages and "formal systems," in which there are axioms and rules for deduction. Mathematics explores what can be deduced from the chosen axioms.

Much fascinating work can be done in studying deductions and proofs. But this study is only one aspect of the whole of mathematics. Formalism by itself does not explain why certain axioms are chosen in preference to others. The axioms that are chosen are ones that are fruitful. The axioms match the world or they match certain pieces of mathematics already done less formally. Formalism does not account for these relationships that extend outside the formal system and are the key reason motivating its study.

Nor does formalism explain the ways that mathematicians search for new results and new theorems. They do not merely manipulate formal symbols according to formal rules. They use intuitive conceptions and pictures that guide them in a search for a more formal proof. Thus, even when the mathematical results are formalized afterward, the formalization captures only one aspect of the whole. It does not deal well with the existential perspective, which includes the intuitions of mathematicians, nor with the situational perspective, which includes the applicability of mathematics to the world.

Gödel's proof has had an effect on formalism as well as on logicism. It established not only that one cannot build mathematics wholly on logic, but also that one cannot build it wholly on a formalization of the axioms within formal language.[8]

Predicativism

Predicativism is a philosophical approach that is more complex, and therefore more difficult to explain in simple terms. It accepts the natural numbers as a given. The natural numbers are given either by our intuition or by a Platonic realm or by both. But predicativism tries to avoid the paradoxes, like Russell's paradox,[9] by being modest about what sets can be constructed using the natural numbers as a base.

For example, predicativism accepts by intuition the set whose members are all natural numbers. It also accepts the set of positive even numbers, because this set can be defined as a subset of the natural numbers, using a clearly defined property ("even"). It does not, however, accept a set that is defined in a way that already implicitly refers to the set in question. Such a definition is called *impredicative*.[10]

The modesty is understandable, given that paradoxes have arisen when people have become overconfident that their intuitions must match the Platonic realm. But predicativism is less a complete philosophy than it is a program recommending a certain kind of modesty. It does not explain the multitude of relationships that we have seen between mental mathematics (the existential perspective), mathematics applied to the world (the situational perspective), and mathematics as a reflection of a transcendent norm (the normative perspective).

[8] Poythress, *Logic*, chapters 55–58 and appendix D1.
[9] Ibid., appendix A1.
[10] Horsten, "Philosophy of Mathematics," §2.4.

Other Philosophical Approaches

We may also mention briefly some other philosophical approaches to mathematics. First, William van Orman Quine advocated a philosophical methodology that came to be called *philosophical naturalism*.[11] He suggested that our best knowledge was from scientific theories, and that philosophy should take its clue from scientific knowledge. In philosophy of mathematics, this approach means that we accept mathematics that is used in the sciences. This approach has the obvious disadvantage that it leaves unexamined the foundations of science.

Another position, *structuralism*, says that mathematics does not describe "entities" in an abstract, Platonic realm, but *structures*, which are characterized by laws and relationships. The natural numbers, for example, are a structure with rules for addition and multiplication, and we can add to these rules more complex relationships (for example, exponents, prime numbers, factorization, numerical representation in base 10 or base 2).

This position has an affinity with the multiperspectival position that we have adopted. There are multiple relations between numbers and the human mind and the world. But by itself structuralism does not explain why some structures with some laws are privileged over others in mathematical studies. So its explanation is still one-dimensional.

Another position, nominalism, tries to dispense with abstract entities like numbers altogether, and to deal only with concrete instantiations, which it enlists to play the roles formerly played by abstract entities. But this position has difficulties. It is beset, to begin with, by the same difficulties that beset medieval nominalism (chapter 2). In addition, it does not account well for the activity of mathematicians, who think about abstractions.

Summary

In addition to the more specialized problems, the secular philosophies share this same great problem: they suppress the revelation of the character of God in mathematics (Rom. 1:18–23).

[11] Ibid., §3.2.

Appendix B

Christian Modifications of Philosophies of Mathematics

Now we consider ways in which Christians have attempted to answer questions in the philosophy of mathematics—questions about the nature of numbers and mathematical objects, and the nature of mathematical truths. There are two main traditional approaches, modified Platonism and modified empiricism.[1]

Christianized Platonism

Christian thinkers who wanted to adopt Plato altered his view by placing Plato's realm of ideas within the mind of God. According to this thinking, Plato's idea of the good exists within God's mind. So does the idea of justice, and the idea of a horse. When applied to arithmetic, this approach implies that numbers and truths about numbers have their original existence in the mind of God.

This Christian alteration of Platonism is an improvement over Plato's own thinking. For one thing, it personalizes truth, by making truths not impersonal abstractions but truths within a personal mind, the mind of God. It also avoids the problem that would be generated if numbers and truths about numbers were eternal realities *independent* of God. If they

[1] James Bradley and Russell Howell, *Mathematics through the Eyes of Faith* (New York: HarperOne, 2011), chapter 10, 221–243. I say "traditional," but modified empiricism, as represented by cosmonomic philosophy, is relatively recent (twentieth century) in comparison with modified Platonism, which goes back to Saint Augustine.

were, it would seem to suggest that they constitute additional absolutes alongside God. They compete with God for ultimacy. But truths within the mind of God obviously do not compete with him.

Christianized Platonism also has a partial answer for the questions about the relation between mathematics as a norm, mathematics as applicable to the world, and mathematics as mental operations in the mind of man. These three correspond to Frame's normative, situational, and existential perspectives, respectively. Secular philosophies of mathematics have deep difficulty in explaining how the three harmonize (appendix A above). Christianized Platonism, on the other hand, can say that they harmonize because God uses the numerical ideas in his mind when he creates the world, thus authorizing the situational perspective. And he makes man in the image of God, with man's mind in harmony with God's mind, and thereby establishes harmony between normative mathematics in God's mind and existential mathematics in man's mind.

However, Christianized Platonism still has difficulties. It suffers from not having dealt fully with the problem of the one and the many (see chapter 2). Christianized Platonism makes the one, namely the original idea in the mind of God, prior to the many, namely the horses or cats or other created things that embody the idea. Likewise, with respect to numbers, Christianized Platonism says that the abstract number 2 within the mind of God is the original idea, and the collections of two apples and two pears in the world are derivative from the idea. The unity of the number 2 is prior to the diversity of collections of two objects in the world.

Let us think about this problem. God's plan for the world exists prior to the world. The world derives from his plan. But his plan includes *both* unity and diversity. He plans to create the species of horse as well as all the individual horses belonging to the species. Likewise, his plan includes both the number 2 and the diversity of collections of two objects. God then executes his plan and brings it into realization in time by creating both the species and some of the individual horses. He executes his plan with respect to numbers by creating a world in which there are collections of two apples and two pears. These collections are "the many." What is common to the collections, namely being collections of two objects, is "the one." So in a more consistently Christian view, the one and the many are equally ultimate.

There are further difficulties with Christian Platonism, concerning its conception of the nature of ideas that it postulates in the mind of God.[2] It does not thoroughly articulate the fact that God is Creator and we are creatures, so that ideas in our mind do not exhaust the ideas in God's mind.

Frame's square on transcendence and immanence, discussed earlier (chapter 5), is relevant. According to a non-Christian view of divine immanence, our ideas, when they are true, are virtually identical to the ideas in God's mind. Our own minds can serve as a standard. While a Christian would naturally deny this principle in most cases, do our minds serve as a final standard in the area of numbers and mathematics? Is our idea of the number 2 identical with the idea in God's mind? How can it be identical without encompassing a knowledge of all the many dimensional relationships between numbers and other things? And if it is not identical to God's idea, is there genuine human knowledge of 2?

Platonism has always suffered from the problem that the kind of knowledge that it postulates must be virtually God-like knowledge of the eternal ideas, if it is to be knowledge at all. Platonism, even Christianized Platonism, runs the danger of breaking down the Creator-creature distinction and falling into non-Christian immanence. Christianized Platonism in mathematics runs the same danger with respect to mathematical ideas and mathematical truths.

Christianized Empiricism

Christianized empiricism is the other major approach to Christian philosophy of mathematics. Christianized empiricism has arisen most prominently with a tradition called *cosmonomic philosophy*. Cosmonomic philosophy is a rich and complex tradition, which we cannot here discuss fully.[3]

We may sketch out the main cosmonomic position briefly, simplifying

[2] Poythress, *Logic*, part I.C.
[3] See Poythress, *Redeeming Philosophy*, appendix A.

at some points. Cosmonomic philosophy, in one of its common forms, says that numbers and truths about numbers are part of the created order. The same principle holds for space and for truths about space. The truths about numbers and space have been *created* and are not eternal. This way of construing numbers conspicuously avoids the difficulties of Platonism, which has to postulate the eternality of numbers and of ideas about space, and thereby runs the danger of producing a second eternality in competition with the eternality of God. In cosmonomic philosophy, there can be no such competition, because only God is eternal, while numbers and space are not. (God knows from eternity what he will create, but that is another matter.)

Cosmonomic philosophy distinguishes two aspects of the created order: (1) created things, such as rocks, plants, animals, and human beings; and (2) laws *governing* the created things. Collections of two apples or two pears are created things. The laws governing collections of two things, such as the law that 2 + 2 = 4, are laws and *not* things. There are many kinds of laws, which govern numbers, space, motion, physical interactions, language, and so on. Together, these laws are the laws of the cosmos—hence the term *cosmonomic*, which comes from two Greek words, for cosmos and for law.

Cosmonomic philosophy is akin to empiricism in its view of numbers, because it maintains that numbers are first of all characteristic of the world around us. We learn from the world what numbers are. But it avoids many of the problems of secular empiricism. It can affirm all three of Frame's perspectives in harmony, because it acknowledges God as Creator. God created the laws concerning numbers (normative perspective); God created the created things that are subject to the laws (situational perspective); and God created human beings in the image of God (existential perspective). Since human beings are made in the image of God, they can faithfully interpret what they see in the world, including what they see concerning its quantitative nature. The same principle goes for space as well as quantity. And from there the principle can be extended to all of mathematics, which represents more complicated forms of law that God ordained for the cosmos.

Cosmonomic philosophy might also give an answer to the problem of the one and the many. God created both aspects of the world. So neither needs to be prior to the other. (This solution is unlike Christianized Platonism, which gives definite priority to the one, the original idea in the mind of God.) On the other hand, a critic might still wonder whether the way in which cosmonomic philosophy gives its description of creation involves some subtle prioritizing of the one to the many. The law appears to be one in relation to the many created things that it governs. Since it *governs* the many, it is in some sense prior to them.

A biblically based approach can affirm that God's speech specifies the whole creation. It specifies *both* the general principles, which express unities, and the individual items and events, which express diversities. But most expositions of cosmonomic philosophy do not appear to have taken this route in discussing *laws* for the cosmos. They treat the law as general, not specifically a law for one collection of two specific apples. But perhaps this is only a superficial preference.

Cosmonomic philosophy has an appeal because it is more "modest" about how we know the mind of God. We as human beings know the mind of God only in the context of the world that God created and in the context of our own finite minds. In fact, we never have a direct divine vision of the ideas in God's mind. Nor do we have a direct vision of numbers as one kind of idea that allegedly exists in God's mind. How, in fact, do we know what the "organization" of God's mind is like? Cosmonomic philosophy would advise us to come down to earth and avoid speculation that is not suitable for us as creatures, who live underneath and not above God's laws.

The difficulty here is the opposite of Christianized Platonism. The difficulty is that we may unwittingly fall into a form of non-Christian transcendence. If we are not careful, we may drift into a form of thinking in which we think that God is unknowable, distant, behind the law, while the law is the only thing that we can actually access. If, for example, the number 3 is created, and not eternal, how can we say that God is three persons? We could say that he appears to us who are underneath the law as three persons. But God who is behind the law cannot really be three persons, because threeness is not eternal. Nor can we understand what it would mean for the Word to be with God eternally, because that involves using a distinction between the Word and God, as two persons, and thus involves the number 2, which allegedly is merely a creation of God, and did not exist eternally.

The people who developed cosmonomic philosophy were believers in Christ, who held orthodox Trinitarian beliefs. I am glad that they did. And I also trust that they genuinely intended that cosmonomic philosophy would be compatible with Trinitarian belief and with the knowability of God, in the sense of a Christian view of immanence, that is, corner #2 of Frame's square. But the philosophy they articulated left this point unclear at best. Its discussion of law is muddied.[4] The muddiness appears to me unfortunately to leave the door open to an interpretation where we actually fall victim to non-Christian thinking about God's transcendence and immanence. And this muddiness affects the philosophy of mathematics, as well as every other area of thought.

In addition, if we fall into a non-Christian view of God's transcendence, we easily also fall victim to a non-Christian view of God's immanence. Let us illustrate how the reasoning could go. Suppose Jill assumes that God is beyond number, because the laws are created. Then she reasons that numbers belong to us as human beings who interact with the cosmos. Since the cosmos is created in a unified way, her own thinking about numbers can, at least apart from the fall, serve as a standard. She realizes that she does not need to claim that she herself is the absolutely ultimate standard that belongs only to God. Rather, the position that she occupies is a position as a proximate standard. It is, however, the only standard that she actually needs within the world, because she has been created by God so that she fully conforms in her thinking to the way the world is. So she can be a master in her thinking in that respect.

Now Jill comes to confront God's revelation to her in Scripture. She reasons (falsely, by the way) that she can still be master, because when the word comes to her, it comes within the cosmic order. Therefore, she reasons, she can use her normal standards for numbers in examining Scripture. And therefore also she concludes that Trinitarian theology is false, because it does not rationally and transparently conform to her preestablished standards.

[4] See ibid.

Thus she has arrived at a non-Christian view of immanence, in which her ideas about number serve as standard, and are used to judge alleged claims of divine revelation.

I should stress that cosmonomic philosophers as orthodox Christian believers did not want any of this train of reasoning. By spinning out Jill's reasoning, I am illustrating the dangers that accrue when we remain unclear about the difference between Christian and non-Christian forms of transcendence and immanence. And these differences affect reasoning about the quantitative order of things.

Thus, we need to maintain the Creator-creature distinction, and to maintain what goes along with it, a Christian view of transcendence and immanence. This view needs to remain in place when we think about mathematics. God's thoughts are superior to ours (Christian transcendence); in addition, his power and his revelation give us genuine access to his thoughts (Christian immanence). The modern world is used to thinking about one level, not two—it ignores the Creator-creature distinction. Breaking with this modern way of thinking can take effort, but is an integral aspect of being faithful followers of Christ.

Appendix C

Deriving Arithmetic

In chapter 9 we introduced Peano's axioms. With those axioms as a starting point, we can define addition and multiplication and derive arithmetical truths. We illustrate the process here with simple beginnings.

Addition

We can define addition using the successor relation, symbolized by S (as explained in chapter 9). In our definitions, the symbols m and n designate natural numbers.

(a) Define $m + 1$ to be Sm, the successor of m: $m + 1 = Sm$. That is, the operation of adding 1 to m has as its result the successor of m.
(b) Define $m + Sn = S(m + n)$. That is, once addition with n has been defined, the operation of adding the successor of n (Sn) to m is defined as the successor of the number obtained by adding n to m.

These two definitions together allow us to define addition for all natural numbers m and n. Why? Because, no matter now large n is, we can gradually reach it by starting with the definition (a) and then repeatedly using (b). The repeated use of (b) in effect uses the principle of mathematical induction. The property M in this case is the following property: M is said to hold true for the number n if the process of adding the number n to other numbers (m) has been defined. The property M clearly holds for $n = 1$, because of definition (a). And definition (b) implies that the property M will always hold for the successor of n, once it holds for n.

$2 + 2 = 4$

Let us show that $2 + 2 = 4$. Now 2 is defined to be the successor of 1: $2 = S1$. Then 3 is defined to be the second successor of 1: $3 = S2 = SS1$. Finally, 4 is defined to be the third successor after 1: $4 = S3 = SS2 = SSS1$.

> $2 + 1 = S1 + 1$ (by definition of 2) $= SS1$ (by definition (a)) $= 3$ (by definition of 3).
>
> $2 + 2 = 2 + S1$ (by definition of 2) $= S(2 + 1)$ (by definition (b)) $= S3$ (by the previous line) $= 4$ (by definition of 4).

If we want to establish results for larger numbers, it just takes more time. Consider, for example, some cases involving addition to the number 3:

1. $3 + 1 = S3$ (by (a)) $= 4$ (by definition of 4).
2. $3 + 2 = 3 + S1 = S(3 + 1)$ (by (b)) $= S4$ (by line 1 above) $= SSSS1$ (by definition of 4) $= 5$ (by definition of 5).
3. $3 + 3 = 3 + SS1$ (by definition of 3) $= S(3 + S1)$ (by (b)) $= S(3 + 2) = S5$ (by line 2) $= 6$ (by definition of 6).
4. $3 + 4 = 3 + SSS1$ (by definition of 4) $= S(3 + SS1)$ (by (b)) $= S(3 + 3)$ (by definition of 3) $= S6$ (by line 3) $= 7$ (by definition of 7).
5. $3 + 5 = 3 + S4 = S(3 + 4)$ (by (b)) $= S7 = 8$ (by definition of 8).

The Associative Law for Addition

Let us try to establish a general result:

> Theorem: for all natural numbers k, m, and n, $(k + m) + n = k + (m + n)$.

This theorem is called the *associative law* for addition.

Let us first try to establish the simpler result that $(k + m) + 1 = k + (m + 1)$. Let us call it a *lemma* (a result that will be used later).

> Lemma: $(k + m) + 1 = k + (m + 1)$.

Proof: We use mathematical induction, where we treat k as fixed, and we try to go through the numbers m starting with $m = 1$. This process is called *induction on m*. In this case, the property M for mathematical induction is the property that $(k + m) + 1 = k + (m + 1)$.

Step (a). Is the principle $(k + m) + 1 = k + (m + 1)$ true when $m = 1$?

> $k + (1 + 1) = k + S1$ (by definition of addition by 1 in definition (a)) = $S(k + 1)$ (by (b)) = $(k + 1) + 1$ (by definition of addition by 1).

Step (b). Given that the principle is true for m, that is, that $(k + m) + 1 = k + (m + 1)$, is it true for $m + 1$?

> $(k + (m + 1)) + 1 = (k + Sm) + 1 = S(k + m) + 1$ (by definition of addition) $= SS(k + m) = S((k + m) + 1) = S(k + (m + 1))$ (by assumption) $= k + S(m + 1)$ (by definition of addition) $= k + ((m + 1) + 1)$.

Now we are ready to try to establish the general principle, $(k + m) + n = k + (m + n)$. We use induction on n, beginning with $n = 1$.

Step (a). Is the principle $(k + m) + n = k + (m + n)$ valid when $n = 1$?

> $(k + m) + 1 = k + (m + 1)$, as just established in the lemma.

Step (b). Assume that the principle is valid for n. Can we establish it for $n + 1$?

> $(k + m) + (n + 1) = ((k + m) + n) + 1$ (by the lemma) $= (k + (m + n)) + 1$ (by assumption) $= S(k + (m + n)) = k + S(m + n)$ (by definition of addition) $= k + (m + Sn)$ (by definition of addition) $= k + (m + (n + 1))$.

So the principle holds for $(n + 1)$. Since we have done both steps (a) and (b), we conclude by mathematical induction that the principle holds for all numbers.

The Commutative Law of Addition

Here is a second theorem:

> Theorem: $m + n = n + m$.

This theorem is called the *commutative law* for addition. Again we can use a lemma:

Lemma: $m + 1 = 1 + m$.

Proof: By induction on m.

Step (a). Is the lemma valid when $m = 1$?

$1 + 1 = 1 + 1$.

Step (b). Assume that the lemma is valid for m. We try to establish it for $m + 1$.

$(m + 1) + 1 = \mathrm{S}(m + 1) = \mathrm{S}(1 + m)$ (by assumption) $= (1 + m) + 1 = 1 + (m + 1)$ (by the associative law).

Now we are ready to prove the general principle of the commutative law, $m + n = n + m$. We do so by induction on n.

Step (a). Is the commutative law valid when $n = 1$?

$m + 1 = 1 + m$, which is just the lemma already proved.

Step (b). Assuming that $m + n = n + m$, can we show it is true for $n + 1$?

$m + (n + 1) = m + \mathrm{S}n = \mathrm{S}(m + n)$ (by definition of addition) $= \mathrm{S}(n + m)$ (by assumption) $= n + \mathrm{S}m$ (by definition of addition) $= n + (m + 1) = n + (1 + m)$ (by lemma) $= (n + 1) + m$ (by the associative law).

Defining Multiplication

In a similar way, we can define multiplication and prove its properties. We will go only a little way in the process.

(a) Define $m \times 1$ to be m: $m \times 1 = m$. That is, the operation of multiplying m by 1 has as its result m itself.
(b) Define $m \times \mathrm{S}n = (m \times n) + m$. That is, once multiplication by n has been defined ($m \times n$), the multiplication by $\mathrm{S}n$ (or $n + 1$) is defined by adding m to the previous result $m \times n$.

The net result of these definitions is that $m \times n$ is the result of adding m to itself for a total of n copies of m.

Using this definition, we can show that $2 \times 2 = 4$.

> $2 \times 2 = 2 \times S1$ (by definition of 2) $= (2 \times 1) + 2$ (by definition of multiplication part (b)) $= 2 + 2$ (by definition of multiplication part (a)) $= 4$.

Conclusion

These exercises may seem tedious. But they show that elementary truths of arithmetic can be derived from simpler principles, namely Peano's axioms. In harmony with our principle of antireductionism (chapter 4), we do not say that this procedure "reduces" numbers to Peano's axioms. Rather, Peano's axioms show one kind of rich relationship between the numbers and between arithmetical truths and logical derivations. We could turn the process on its head, and say that Peano's axioms are "derived" from the truths of arithmetic by selecting certain truths. In order later to serve as axioms, the truths that we select must together be enough to derive the rest.

Appendix D

Mathematical Induction

We include other illustrations of mathematical induction.

Sum of Odd Numbers

The sum of the first n odd numbers is $n \times n = n^2$. We can check the truth of this claim by testing the first few cases:

$1 = 1 \times 1 = 1^2$.

$1 + 3 = 4 = 2 \times 2 = 2^2$.

$1 + 3 + 5 = 9 = 3 \times 3 = 3^2$.

$1 + 3 + 5 + 7 = 16 = 4 \times 4 = 4^2$.

But how would we check the truth for *every* case? We can never complete the process. This kind of situation makes plain the value of mathematical induction.

We proceed by induction on n. Step (a) consists in checking the truth for the value $n = 1$. We have already done that above: $1 = 1 \times 1 = 1^2$.

Step (b) begins by assuming that the principle is valid for the number n, and then trying to establish it for $n + 1$. Assume that

$$1 + 3 + \ldots + (2n - 1) = n^2.$$

Now try to do the next case, for $n + 1$.

$$1 + 3 + \ldots + (2n - 1) + (2n + 1) = ?$$

Since the first n terms in this sum are the same as in the previous equation, we can substitute n^2 for the sum of the first n terms:

$$1 + 3 + \dots + (2n - 1) + (2n + 1) = n^2 + (2n + 1) = n^2 + 2n + 1 = (n + 1)^2.$$

This shows that the formula holds for $n + 1$. So, by the principle of mathematical induction, the formula holds for all n whatsoever.

The principle of mathematical induction enables us to avoid having to do an infinite number of distinct calculations. We can understand and use this principle because we are made in the image of God, and we can transcend the particularities of an individual calculation in order to understand the general pattern.

Using the result for the sum of odd numbers, we have a wonderfully easy way to calculate the sum of even numbers.

The sum of the first n even numbers is $n^2 + n$.

We could establish this formula by using induction on n. But there is a simpler way of doing it. Consider again the sum of the first n odd numbers:

$$1 + 3 + 5 + \dots + (2n - 1) = n^2.$$

Now add a 1 to each of these n numbers:

$$(1 + 1) + (3 + 1) + (5 + 1) + \dots + ((2n - 1) + 1) \text{ or}$$
$$2 \quad + \quad 4 \quad + \quad 6 \quad + \dots + \quad 2n$$

The result is the sum of the first n even numbers. Since we have arrived at this sum by adding a total of n 1's to the original sum, which was n^2, the sum of the first n even numbers must be the sum n^2 of the first n odd numbers, plus an additional n, for a total of $n^2 + n$. So

$$2 + 4 + 6 + \dots + 2n = n^2 + n.$$

If we take the sum of the first n even numbers, and divide term by term by 2, we obtain:

$$2/2 + 4/2 + 6/2 + \ldots + 2n/2 = 1 + 2 + 3 + \ldots + n = (n^2 + n)/2 = n(n + 1)/2.$$

That is, the sum of the first n numbers is $n(n + 1)/2$. This result could also be obtained directly by mathematical induction.[1] God has ordained marvelous harmonies by providing several ways in which the same results may be checked out.

The Sum of Squares

The sum of the squares of the first n numbers is $n(n + 1)(2n + 1)/6$. Again, we can verify the first few cases:

$$1^2 = 1 = 1(1 + 1)(2 \times 1 + 1)/6.$$
$$1^2 + 2^2 = 1 + 4 = 5 = 2(2 + 1)(2 \times 2 + 1)/6.$$
$$1^2 + 2^2 + 3^2 = 1 + 4 + 9 = 14 = 3(3 + 1)(2 \times 3 + 1)/6.$$
$$1^2 + 2^2 + 3^2 + 4^2 = 1 + 4 + 9 + 16 = 30 = 4(4 + 1)(2 \times 4 + 1)/6.$$

Now let us try to show it is always true.

Step (a). Show that it is true for $n = 1$. We have already shown it above.
Step (b). Assume that it is true for n. That is, assume that

$$1^2 + 2^2 + 3^2 + \ldots + n^2 = n(n + 1)(2n + 1)/6.$$

The sum for $n + 1$ is

$$1^2 + 2^2 + 3^2 + \ldots + n^2 + (n + 1)^2 = n(n + 1)(2n + 1)/6 + (n + 1)^2$$
(by assumption that the formula holds for n).

Regrouping,

$$n(n + 1)(2n + 1)/6 + (n + 1)^2 = [n(n + 1)(2n + 1) + 6(n + 1)^2]/6$$
$$= (n + 1)[n(2n + 1) + 6(n + 1)]/6 = (n + 1)[2n^2 + n + 6n + 6]/6$$
$$= (n + 1)(n + 2)(2n + 3)/6 = (n + 1)((n + 1) + 1)(2(n + 1) + 1)/6,$$

which confirms that the formula holds for $n + 1$. By induction, it holds for all n.

---

[1] See Poythress, *Redeeming Science*, appendix 2.

More Perspectives on the Sum of Odd Numbers

We can use other perspectives to show that the sum of the first n odd numbers is n^2. Here is the sum:

$$1 + 3 + 5 + \ldots + (2n - 3) + (2n - 1)$$

Write the same sum in the reverse order, and put this new sum directly under the first:

$$\begin{array}{ccccccccccc} 1 & + & 3 & + & 5 & + \ldots + & (2n-3) & + & (2n-1) \\ (2n-1) & + & (2n-3) & + & (2n-5) & + \ldots + & 3 & + & 1 \end{array}$$

Now add the two lines, term by term:

$$2n \quad + \quad 2n \quad + \quad 2n \quad + \ldots + \quad 2n \quad + \quad 2n$$

Since we started with n odd numbers, there are n copies of $2n$, for a total of $2n^2$. This total is the result of adding the original sum of n odd numbers to itself. So the sum of n odd numbers is half of $2n^2$, or n^2.

A second perspective on the same sum uses a pictorial diagram to enable us to see the arithmetical truth (diagram D.1).

In the diagram, the L-shaped regions all contain an odd number of dots. The odd numbers add up to make a square region containing a number of dots that is a square number. For example, the number of dots in the square region with 5 dots on a side is 5^2. Inspecting the diagram shows us that the same number of dots is also $1 + 3 + 5 + 7 + 9 = 5^2$.

A third perspective on the same problem focuses on the difference between n^2 and the next square, $(n + 1)^2$. $(n + 1)^2$, when multiplied out, is the same as $n^2 + 2n + 1$. Hence,

$$(n + 1)^2 - n^2 = 2n + 1,$$

which is an odd number. As n increases, the differences between the squares are the successive odd numbers. If we start with $n = 1$, we obtain the result that $2^2 - 1^2 = 3$. If we arrange the squares in a row, with their differences below, we obtain diagram D.2. Each square is 1 (the first square) plus the sum of all the differences to its left and below it. This diagram enables us to take in at a glance the result from mathematical induction, where we assume the truth for n and try to establish it for $n + 1$.

Diagram D.1: Dots in a Square

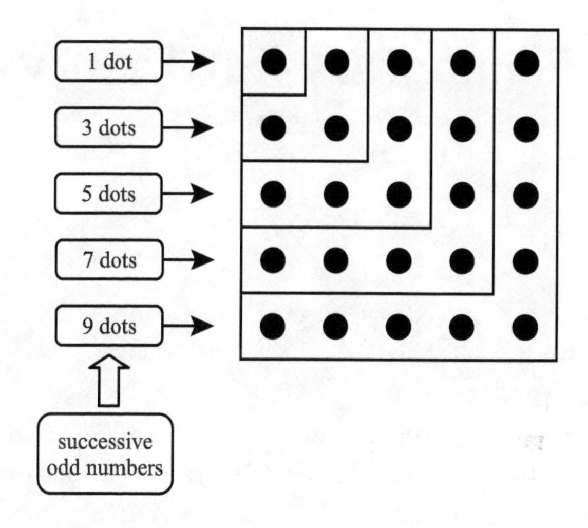

Diagram D.2: Squares and Differences

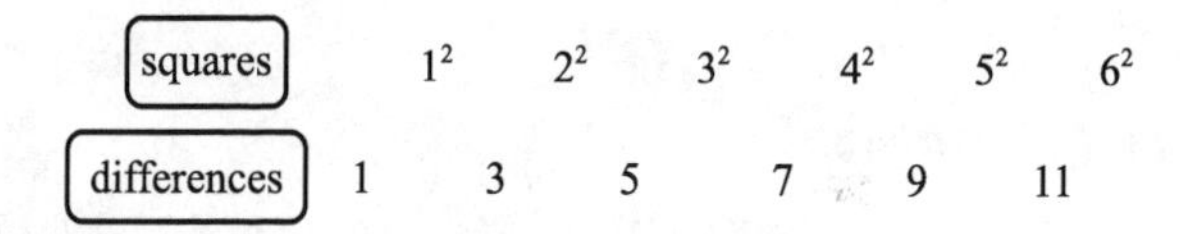

The multiple perspectives for looking at the sum of odd numbers show the richness of the truths that God has ordained.

Appendix E

Elementary Set Theory

In chapter 12 we indicated that sets can be used to introduce axioms from which the truths of arithmetic can be derived. We would like to explore the first steps in this process.

There are several possibilities for starting axioms. The most common starting point has become *Zermelo-Fraenkel* set theory, the work of Ernst Zermelo and Abraham Fraenkel in the early twentieth century.[1] We will set forth some of their axioms. But when possible we will express the meaning of the axioms in ordinary English, so that even readers without a mathematical background can understand the central point.

The Axiom of Extension

The *axiom of extension* says that two sets are identical if they contain the same elements.

This axiom indicates that the concept of set is a "stripped down" concept. We ignore every kind of information except the specifications for which elements belong to the set.

The axiom depends on our ability as human beings made in the image of God to see the general pattern common to many concrete collections, and to produce a concept that focuses only on what is common. For example, if we have two apples sitting on a table, we can think about the apples in more than one way. If we focus on their past, we can observe

[1] Thomas Jech, "Set Theory," *The Stanford Encyclopedia of Philosophy (Winter 2011 Edition)*, ed. Edward N. Zalta, http://plato.stanford.edu/archives/win2011/entries/set-theory/, §4, accessed June 18, 2014.

that they came from the same bag bought at the grocery store. Or they were picked from the same tree. If we focus on the present, we can observe that they are both on the table. If we focus on their future, we can observe that they are going to be eaten as part of a salad or part of an apple pie. In terms of full meaning, we can have several "collections"—the collection of apples from the same bag, or the collection of apples on the table, or the collection of apples that will make up the same pie. If, on the other hand, we strip away the extra meaning, we have one "set" of two apples. The "set," in the technical sense, depends *only* on what things are its members, not on any extra knowledge about the members and why they are considered as part of a single collection.

The concept of a set also depends on our knowledge of what it means to be "the same" element. For example, we have to be able to identify an apple as the "same" apple, even though it ripens over time. As we indicated in chapter 12, the concept of "being the same element" already uses the idea of unity in diversity. The unity is the unity of the "same" apple. The diversity—or at least one kind of diversity—is visible in the ripening process over time. In this use of unity and diversity, we already depend on our understanding of numbers. The unity is the unity of one thing, the diversity is the diversity of more than one phase of the thing. We also rely on our pre-theoretical understanding of collections. We must understand tacitly what it means to have two apples on the table, and mentally to consider them as belonging together.

The Axiom of the Null Set

The *axiom of the null set* says that there exists a set with no members. This set is conventionally designated $\emptyset$, and is also called the *empty set*.

This axiom is not so intuitive. Is a "set" with no members really a set? Or is it nothing? The conception of a null set is analogous to the conception of the number zero. Is the number zero a number? Or is it nothing? In a way the decision is up to us—it depends on how expansive we want to make our own conception of "set." We can understand the concept of the null set more intuitively if we think about the process of "subtracting away" one member from a set that has more than one member to begin with. Suppose we begin with a set with two members: $\{2, 3\}$. If we omit 3 as a member, we get a second set, the set whose only member is 2: $\{2\}$. It seems reasonable to allow that we might get a third set by omitting 2 as well, in which case we the set $\{\}$ with no members. (The symbol $\emptyset$ is customarily used instead of the symbol $\{\}$, but this is merely a matter of notation.)

We see an analog to this concept of a null set in the account of creation in Genesis 1. Genesis 1 describes the initial situation as one where "the earth was without form and void (empty)" (v. 2). The context indicates that some things were present: the earth itself, the waters, and the Spirit of God. But the earth was empty of plants and animals, the kind of discrete furnishings that it would later enjoy. The language in Genesis is ordinary language, not the technical language of mathematics. But it implies that God has designed a world where the concept of a collection of furnishings for the world is appropriate. And at an early point in time this collection of furnishings was empty. God, in making us in his image, gave us the capability of thinking in terms of a null set.

The Axiom of Pairs

The *axiom of pairs* says that if we have two elements or sets a and b, there is a set whose members are a and b: $\{a, b\}$.

This axiom seems to be so simple that one might wonder why it is included at all. One of the reasons for having axioms is to make all the assumptions explicit, and leave nothing that is being used as an extra *implicit* assumption. The axiom of pairs means that if we already have some sets a and b, we can build more. One of the consequences is that if a is a set, $\{a, a\}$ is a set. But $\{a, a\}$ is the same as $\{a\}$ (since it has only one member, a). Is a the same as $\{a\}$? No. $\{a\}$ has one member, namely a. If a is a set, it may have many members or none at all (it may be the null set). So in general $a \neq \{a\}$.

God gives us the ability to make distinctions. Among the distinctions we may make is one where we distinguish two items a and b from all the other possible items. We are presupposing our capacity to make distinctions. We also presuppose that, after making a distinction, we can have a group of items that are "inside" and distinguished from all the rest of the world. In addition, the capacity to group together two items, after we already have one, displays an instance of additivity. We are really presupposing the idea of addition, which goes back to God.

The Axiom of Subsets

The *axiom of subsets* says that if we have a set A, there is a set B that consists in all the members of A that have a specific additional property (besides being in A).

For example, suppose that A is the set of odd numbers less than 10: $A = \{1, 3, 5, 7, 9\}$. Let B consist in those members of A that are perfect squares. $1 = 1 \times 1$, so 1 is a perfect square. $9 = 3 \times 3$, so 9 is a perfect square. The other numbers 3, 5, and 7 are not perfect squares. So $B = \{1, 9\}$. The axiom of subsets says that if the set A exists, B also exists.

This axiom depends on our understanding of how to separate out *some* but not all of the members of a set by specifying an additional property. The additional property separates some elements from the rest. It is a distinction. The idea of distinction, as we have seen, has its roots in God (chapter 12).

The Axiom of the Sum Set

Suppose we start with a set A. The *axiom of the sum set* says that we can "sum up" all the elements that are members of all the members of A, and make a single set that has all of them as its members.

This idea can be confusing. So let us consider an example. Let the set A have as its members several other sets B, C, and D. That is, $A = \{B, C, D\}$. The sum set of A is the set U consisting of all elements that are members of B or C or D (including elements that are members of more than one of these three). That is,

$$U = \{x \mid x \in B \text{ or } x \in C \text{ or } x \in D\}.$$

This sum set of A is denoted $\cup A = U$. So

$$\cup A = U = \{x \mid x \in B \text{ or } x \in C \text{ or } x \in D\} = \{x \mid \text{for some } y, x \in y \in A\}$$

The symbol $\cup$ is also used in another, related way to indicate the union, $B \cup C$, of two sets B and C. The union is the set whose members are the elements that are either in B or in C or in both. If B and C are both sets, the axiom of pairs says that $\{B, C\}$ is a set. Then the axiom of the sum set says that $\cup\{B, C\}$ exists. $\cup\{B, C\}$ is the same as $B \cup C$.

The axiom of the sum set says that we can collect together all the members from a list of sets. We creatively form a new collection. This creativity is an image of divine creativity. This axiom together with other axioms allows us to form sets with more and more members. In doing so, we use the idea of what is next. In particular, we can produce a series of sets that together mimic the series of natural numbers. Let us see how.

Producing a Sequence of Sets

Corresponding to the number zero we will have the empty set $\emptyset$. The empty set exists, according to the axiom of the null set. It has zero elements. We will start with the number zero rather than with the number one, so that later on each number n will correspond to a set with exactly n elements. The empty set, with 0 elements, corresponds to the number zero.

Since the empty set exists, the axiom of pairs implies that the set $\{\emptyset, \emptyset\}$ exists. Since the element $\emptyset$ is identical to itself, the set $\{\emptyset, \emptyset\}$ is the same as $\{\emptyset\}$. It has one element in it, namely $\emptyset$. To this set will correspond the number 1.

Because $\emptyset$ and $\{\emptyset\}$ both exist, the axiom of pairs implies that the set $\{\emptyset, \{\emptyset\}\}$ exists. It is the set with two elements $\emptyset$ and $\{\emptyset\}$. The two elements are not identical, since $\emptyset$ has no members and $\{\emptyset\}$ has one member (namely $\emptyset$). It is obvious at this stage that we have to distinguish carefully between a set and the elements that are its members. The set $\{\emptyset, \{\emptyset\}\}$ will correspond to the number 2.

Since $\{\emptyset\}$ and $\{\emptyset, \{\emptyset\}\}$ both exist, the axiom of pairs says that there exists a set

$$\{\{\emptyset\}, \{\emptyset, \{\emptyset\}\}\}.$$

Again using the axiom of pairs, there exists a set

$$\{\{\emptyset\}, \{\{\emptyset\}, \{\emptyset, \{\emptyset\}\}\}\}.$$

The axiom of the sum set, applied to $\{\{\emptyset\}, \{\{\emptyset\}, \{\emptyset, \{\emptyset\}\}\}\}$, implies the existence of

$$\cup\{\{\emptyset\}, \{\{\emptyset\}, \{\emptyset, \{\emptyset\}\}\}\} = \{\emptyset, \{\emptyset\}, \{\emptyset, \{\emptyset\}\}\}$$

This set has three members, $\emptyset$, $\{\emptyset\}$, and $\{\emptyset, \{\emptyset\}\}$. It will correspond to the number 3.

This process has been laborious, but we have produced sets with 0, 1, 2, and 3 elements, respectively. We could continue, and produce sets with even more elements. To see the point quickly, we can adopt the option, sometimes used in set theory, of actually using sets as the *names* of numbers (or numbers as the names for some sets, which amounts to the same thing). For the purposes of the theory, a number is *identified* with the set. So, for example, the number 0 becomes an abbreviation for the empty set $\emptyset$: $0 = \emptyset$. The number 1 becomes an abbreviation for $\{\emptyset\}$: $1 = \{\emptyset\} = \{0\}$. The number 2 becomes an abbreviation for $\{\emptyset, \{\emptyset\}\}$: $2 = \{\emptyset, \{\emptyset\}\} = \{0, 1\}$. And the number 3 becomes an abbreviation for $\{\emptyset, \{\emptyset\}, \{\emptyset, \{\emptyset\}\}\}$: $3 = \{\emptyset, \{\emptyset\}, \{\emptyset, \{\emptyset\}\}\} = \{0, 1, 2\}$. This notation enables us easily to see a pattern. We can continue the pattern: $4 = \{0, 1, 2, 3\}$; $5 = \{0, 1, 2, 3, 4\}$; $6 = \{0, 1, 2, 3, 4, 5\}$. We can see (by imitative transcendence) that the same pattern enables us to produce a number as large as we want. We can extend the sequence indefinitely.

In general, if we use this convention for numbers, the successor of the number n is $n \cup \{n\}$. Once we have a successor relation, we can proceed to define addition and multiplication as in appendix C. But as an axiom we need also to include some form of the principle of mathematical induction.

The Axiom of Infinity

So far, we have been able to build sets with a finite number of elements. To obtain resources for arithmetic, set theory needs an *axiom of infinity*. This axiom can take the form of saying that there exists a set that includes as members all the sets 0, 1, 2, 3, … . The minimal set with this property is the set of nonnegative integers, conventionally designated ℕ. (In the context of set theory, it is also designated ω.)[2]

Chapters 8 and 9 indicated how we rely on God for the idea of an indefinitely extended sequence. The infinity of God is the ultimate foundation for our ability to think of an indefinitely extended sequence—an infinite sequence. The same observations apply here. Whether we think directly in terms of numbers or we think in terms of sets that we will use to represent numbers, the same resources are needed, and these resources have their ultimate foundation in God.

The Axiom of Power Set

Zermelo-Fraenkel set theory includes other axioms, which come into play primarily in producing larger sets that are useful in the theory of real numbers and in advanced set theory. The first such axiom is called the *axiom of power set*. To understand it, we should first define the meaning of *subset*. A *subset* of a set *A* is a set *B* all of whose members are members of *A*. So, for example, {1, 3} is a subset of {1, 2, 3}.

The axiom of power set says that, if we have a set *A*, there exists another set, the *power set* of *A*, whose members are all the subsets of *A*. For example, if *A* is {1, 2, 3}, the power set of *A* is the set of all subsets of *A* or {∅, {1}, {2}, {3}, {1, 2}, {1, 3}, {2, 3}, {1, 2, 3}}. Conventionally, the power set of *A* is denoted *P*(*A*). By repeatedly applying the axiom of power set, one can produce very large sets quickly. The power set of the set {1, 2, 3} with 3 members has $2^3 = 8$ members. The power set of a set with 8 members has $2^8 = 256$ members. The power set of a set with 256 members has 2^{256} members, which is approximately 10^{77}, 1 followed by 77 zeros.

The idea of power set shows another use of human power for imitative transcendence (see chapters 8 and 9). When we have a set *A*, we stand back from it and imagine ourselves collecting elements together into a new, more extended set consisting of all the subsets of *A*. We "rise above" the set *A* in the process. We imitate God, whose view of all things is comprehensive, and who rises above them in his infinity.

[2] The symbol ℕ is unicode U2115. ω, the last letter of the Greek alphabet, is unicode U03C9.

The Axiom of Replacement

The *axiom of replacement* says roughly that if we have a set A, and we have a way of correlating each member x in A with a unique set B_x, there is a set whose members are all the sets B_x. This axiom is called the axiom of replacement because the basic idea is to "replace" each member x in A with the correlated set B_x. The result of the replacement is a new set.

Here is an example. Let $A = \{1, 2, 3, 4\}$. Let the numbers 1, 2, 3, and 4 be correlated respectively to the sets {1}, {1, 2}, {1, 2, 3, 4, 5}, and {1, 2, 3, 4, 5, 6, 7, 8}. Then a set exists that has the "replacements" instead of the original members 1, 2, 3, and 4 as its members. The new set has as its members {1}, {1, 2}, {1, 2, 3, 4, 5}, and {1, 2, 3, 4, 5, 6, 7, 8}; that is, it is the set {{1}, {1, 2}, {1, 2, 3, 4, 5}, {1, 2, 3, 4, 5, 6, 7, 8}}.

The axiom of replacement presupposes our ability to make these correlations, and to envision a second, new set with the correlated items as its members. The correspondence between the members of A and the other sets can be viewed as a kind of symmetry, depending ultimately on the original symmetry in God. We are thinking God's thoughts after him, in a complex way.

When we use an example like this, it may not seem too impressive. But when the axiom of replacement is used in connection with the other axioms, it can lead to new sets that are larger than any set produced by other means, because the sets correlated with the members of A can be very large.

Let us consider one example. Begin with the set of nonnegative integers, designated ω. Using the axiom of power set, produce successive power sets of ω: ω, $P(\omega)$, $P(P(\omega))$, $P(P(P(\omega)))$, and so on. Is there a set larger than all the sets in this list? Without the axiom of replacement, we cannot guarantee that there will be a set of the form

$$\{\omega, P(\omega), P(P(\omega)), P(P(P(\omega))), \dots\}.$$

Now we simply correlate 0 with ω, 1 with $P(\omega)$, 2 with $P(P(\omega))$, 3 with $P(P(P(\omega)))$, and so on. Using this correlation and the fact that the set ω ($= \{0, 1, 2, 3, \dots\}$) exists, the axiom of replacement allows us to conclude that $\{\omega, P(\omega), P(P(\omega)), P(P(P(\omega))), \dots\}$ exists. Designate this new set as M. This new set has only as many members as there are nonnegative integers. But the sum set $\cup M$ is very large, including as it does all the subsets of all the power sets in the sequence beginning with ω. Once we have the large set $\cup M$, the axiom of power set allows us to conclude that $P(\cup M)$, $P(P(\cup M))$, $P(P(P(\cup M)))$, … all exist. Since we can correlate 0, 1, 2, 3, … with the sequence $\cup M$, $P(\cup M)$, $P(P(\cup M))$, … , there is a new set whose members are the entire sequence: $\{\cup M, P(\cup M), P(P(\cup M)), \dots\}$.

We can observe in this process the repeated use of imitative transcendence. At each step where we produce larger sets, we go beyond or "transcend" the position at which we have already arrived.

The Axiom of Choice

The final axiom we will discuss is called the *axiom of choice*. The axiom of choice was not part of the original list of axioms proposed by Zermelo and Fraenkel. It is subject to more debate, and some philosophers and mathematicians have expressed uneasiness about it. The intuitionists reject it.

The *axiom of choice* says roughly that, if we have a set A whose members are nonempty sets, we can find a way of picking one designated element out of each set that is a member of A. This axiom may sound trivial. If the member sets are nonempty, it means that each of them has at least one member, and we just pick one. But if we are dealing with an infinite set A, such as the set of natural numbers, we can never complete the process. The axiom of choice is not a logical consequence of the other axioms. But it seems reasonable. Why? We are again using our capacity for imitative transcendence. Even though we can never in practice complete the process of picking an element from each set among an infinite number of sets, we can imagine it being done. We extrapolate to infinity, as it were. Even though we are finite, we are imaginative imitators of infinity. We have an idea of infinity. We do because we are made in God's image and we are imitating him.

Bibliography

Anderson, Stephen R. *Doctor Dolittle's Delusion: Animals and the Uniqueness of Human Language*. New Haven, CT: Yale University Press, 2004.

Beth, Evert W. *The Foundations of Mathematics: A Study in the Philosophy of Science*. 2nd rev. ed. Amsterdam: North-Holland, 1968.

Bishop, Steve. "A Bibliography for a Christian Approach to Mathematics" (June 7, 2008). http://www.scribd.com/doc/3268416/A-bibliography-for-a-Christian-approach-to-mathematics. Accessed September 17, 2012.

Bradley, James, and Russell Howell. *Mathematics through the Eyes of Faith*. New York: HarperOne, 2011.

Byl, John. *The Divine Challenge: On Matter, Mind, Math, and Meaning*. Edinburgh/Carlisle, PA: Banner of Truth, 2004.

Chase, Gene, and Calvin Jongsma. "Bibliography of Christianity and Mathematics, 1st edition 1983." http://www.asa3.org/ASA/topics/Mathematics/1983Bibliography.html. Accessed July 30, 2012. This bibliography was published by Dordt College Press in 1983, but is now out of print.

Frame, John M. *Apologetics to the Glory of God: An Introduction*. Phillipsburg, NJ: Presbyterian & Reformed, 1994.

———. *The Doctrine of God*. Phillipsburg, NJ: Presbyterian & Reformed, 2002.

———. *The Doctrine of the Christian Life*. Phillipsburg, NJ: Presbyterian & Reformed, 2008.

———. *The Doctrine of the Knowledge of God*. Phillipsburg, NJ: Presbyterian & Reformed, 1987.

———. *The Doctrine of the Word of God*. Phillipsburg, NJ: Presbyterian & Reformed, 2010.

———. *Perspectives on the Word of God: An Introduction to Christian Ethics*. Eugene, OR: Wipf & Stock, 1999.

Horsten, Leon. "Philosophy of Mathematics." *The Stanford Encyclopedia of Philosophy (Spring 2014 Edition)*. Edited by Edward N. Zalta. http://plato.stanford.edu/archives/spr2014/entries/philosophy-mathematics/. Accessed June 18, 2014.

Howell, Russell W., and W. James Bradley, eds. *Mathematics in a Postmodern Age: A Christian Perspective*. Grand Rapids, MI/Cambridge: Eerdmans, 2001.

Iemhoff, Rosalie. "Intuitionism in the Philosophy of Mathematics." *The Stanford Encyclopedia of Philosophy (Spring 2014 Edition)*. Edited by Edward N. Zalta. http://plato.stanford.edu/archives/spr2014/entries/intuitionism/. Accessed June 18, 2014.

Jech, Thomas. "Set Theory." *The Stanford Encyclopedia of Philosophy (Winter 2011 Edition)*. Edited by Edward N. Zalta. http://plato.stanford.edu/archives/win2011/entries/set-theory/. Accessed June 18, 2014.

Kuyk, Willem. *Complementarity in Mathematics: A First Introduction to the Foundations of Mathematics and Its History*. Dordrecht-Holland/Boston: Reidel, 1977.

Kuyper, Abraham. *Lectures on Calvinism: Six Lectures Delivered at Princeton University under Auspices of the L. P. Stone Foundation*. Grand Rapids, MI: Eerdmans, 1931.

Meek, Esther L. *Longing to Know: The Philosophy of Knowledge for Ordinary People*. Grand Rapids, MI: Brazos, 2003.

Milbank, John. *The Word Made Strange: Theology, Language, Culture*. Oxford: Blackwell, 1997.

Nickel, James. *Mathematics: Is God Silent?* Rev. ed. Vallecito, CA: Ross, 2001.

Poythress, Vern S. "A Biblical View of Mathematics." In *Foundations of Christian Scholarship: Essays in the Van Til Perspective*. Edited by Gary North. Vallecito, CA: Ross, 1976. Pp. 158–188. http://www.frame-poythress.org/a-biblical-view-of-mathematics/. Accessed December 29, 2012.

———. *God-Centered Biblical Interpretation*. Phillipsburg, NJ: Presbyterian & Reformed, 1999.

———. *Inerrancy and Worldview: Answering Modern Challenges to the Bible*. Wheaton, IL: Crossway, 2012.

———. *Logic: A God-Centered Approach to the Foundation of Western Thought*. Wheaton, IL: Crossway, 2013.

———. "Mathematics as Rhyme." *Journal of the American Scientific Affiliation* 35/4 (1983): 196–203.

———. "Newton's Laws as Allegory." *Journal of the American Scientific Affiliation* 35/3 (1983): 156–161. http://www.frame-poythress.org/newtons-laws-as-allegory/. Accessed June 18, 2014.

———. *Redeeming Philosophy: A God-Centered Approach to the Big Questions*. Wheaton, IL: Crossway, 2014.

———. *Redeeming Science: A God-Centered Approach*. Wheaton, IL: Crossway, 2006.

———. "Science as Allegory." *Journal of the American Scientific Affiliation* 35/2 (1983): 65–71. http://www.frame-poythress.org/science-as-allegory/. Accessed June 18, 2014.

———. *The Shadow of Christ in the Law of Moses*. Phillipsburg, NJ: Presbyterian & Reformed, 1995.

———. *Symphonic Theology: The Validity of Multiple Perspectives in Theology*. Reprint. Phillipsburg, NJ: Presbyterian & Reformed, 2001.

———. "Tagmemic Analysis of Elementary Algebra." *Semiotica* 17/2 (1976): 131–151.

Sayers, Dorothy. *The Mind of the Maker*. New York: Harcourt, Brace, 1941.

Strauss, D. F. M. "The Concept of Number: Multiplicity and Succession between Cardinality and Ordinality," *South African Journal of Philosophy* 25/1 (2006): 27–47, http://www.freewebs.com/dfmstrauss/Ordinality_and_Cardinality.pdf, accessed August 5, 2014.

———. "Frege's Attack on 'Abstraction' and His Defense of the 'Applicability' of Arithmetic (as Part of Logic)," *South African Journal of Philosophy* 22/1 (2003): 63–80.

———. "Infinity and Continuity: The Mutual Dependence and Distinctness of *Multiplicity* and *Wholeness*," paper presented at the Free University of

Brussels, October 15, 2006. http://www.reformationalpublishingproject.com/rpp/docs/Infinity_and_Continuity.pdf, accessed August 5, 2014.

———. "The Significance of Non-Reductionist Ontology for the Discipline of Mathematics: A Historical and Systematic Analysis," *Axiomathes* 20 (2010): 19–52. DOI 10.1007/s10516-009-9080-5, http://link.springer.com/article/10.1007%2Fs10516-009-9080-5#page-1, accessed August 5, 2014.

———. "What Is a Line?" *Axiomathes* 24/2 (2014): 181–205.

van Atten, Mark. "Luitzen Egbertus Jan Brouwer." *The Stanford Encyclopedia of Philosophy (Summer 2011 Edition)*. Edited by Edward N. Zalta. http://plato.stanford.edu/archives/sum2011/entries/brouwer/. Accessed June 18, 2014.

Van Til, Cornelius. *The Defense of the Faith*. 2nd ed., rev. and abridged. Philadelphia: Presbyterian & Reformed, 1963.

Vollenhoven, Dirk H. Theodoor. "Problemen en richtingen in de wijsbegeerte der wiskunde" [Problems and Directions in the Philosophy of Mathematics]. *Philosophia Reformata* 1 (1936): 162–187.

———. *De wijsbegeerte der wiskunde van theïstisch standpunt* [The Philosophy of Mathematics from a Theistic Standpoint]. Amsterdam: Van Soest, 1918.

Whitehead, Alfred North, and Bertrand Russell. *Principia Mathematica*. 2nd ed. 3 vols. Cambridge: Cambridge University Press, 1927.

General Index

Scripture Index